MW01349439

The Underlying Cause of the Unconscious Conspiracy Against Our Health

E. Wil Spencer

No seeds, nuts, grains,
chocolet

The Body Electrician

www.BodyElectrician.com

First Printing: 2013

ISBN 978-1-329-90239-8

The Body Electrician OR Environotics Unlimited

2430 Butler St. Suite 105, Easton, PA 18042

610-417-7248

www.bodyelectrician.com

Dedication

This book is dedicated to my children— Isac, Shoshana, Ama, and Sam—who have each and every time, so lovingly welcomed me home from the office and my trips, and accepted this life that we have, as normal.

Thank you and I love you!

Claimer

The contents of this book are for informational purposes only and do not render medical or psychological advice, opinion, diagnosis, or treatment. The information provided through this book should not be used for diagnosing or treating a health problem or disease. It is not a substitute for professional care. If you have or suspect you may have a medical or psychological problem, you should consult the appropriate health care provider.

Notice

Prelude

Wil Spencer puts his life experience in his first book that takes the twisted line of the question mark (?) left at the end of my life's work and books, to straighten it into an exclamation point (!). What Wil reveals just on what they did to Candida yeast alone is worth the price of the book. When coupled with his views on the dynamic balance [imbalance] of pathogens and genetically modified organisms with Nature, his ideas are paradigm-breaking.

I come from the Illinois farm culture where the response to any and all problems is: "Get a Bigger Hammer". The world is ill. The Bigger Hammer isn't working. I have had to confront that method as the definition of Insanity: To do the same thing every time expecting different results.

If we are to view the Universe mechanistically, then we must accept that there might be a different and better way to approach this imbalance. Wil instructs us in the already established concepts of Bechamps vs. Pasteur: If you want to remove a pathogen - change its environment so that it no longer favors its growth.

It is time to put down our hammers and pick up Wil's book.

Patrick Jordan, Author

ICD-999

Vaccine Induced Diseases

The Chronic Serum Sickness Postulate

www.vaccinefraud.com

Table of Contents

Dedication **3**

Claimer **4**

Notice **4**

Prelude **5**

Forward **11**

Prologue **17**

Preface **19**

Chapter 1 **25**

The Unconscious Conspiracy ***25***

Finding the Simple Solution ***26***

Natural Laws ***28***

Life Originates in the Soil ***29***

A Healthy GI Tract ***31***

Basic Ingredients for Health ***33***

Chapter 2 **37**

Grains, Nuts, Seeds, and Tubers: An Electrical and Biological Nightmare ***37***

Anti-Nutrients ***37***

High Starch and Insoluble Fiber ***39***

Historical Truth ***42***

Chapter 3 **45**

Lactose Intolerance ***45***

Symptom Relief Is Not Cure ***47***

Chapter 4 **49**
The Candida Story *49*
Biological Disaster *50*
Bacillus Laterosporus *52*
Chapter 5 **55**
Candidiasis *55*

Chapter 6 **59**
Candida Symptoms *59*
Allergies and Environmental Sensitivities *59*
Childhood Developmental Concerns *61*
Fatigue *61*
Hormonal Imbalances *62*
Hypoglycemia and Diabetes *63*
Neurological and Emotional Symptoms *64*
Poor Absorption and Elimination *64*
Reproduction, Respiratory and Skin Conditions *66*
Chapter 7 **67**
Do You Have A Candida Overgrowth? *67*

Chapter 8 **69**
How Can I Tell If I Have a Candida Overgrowth? *69*
Simple Saliva Test *69*
Overgrowth Symptoms *71*
Chapter 9 **73**
Simple Program *73*

Chapter 10 **77**
Dietary Parameters of Workability for Maximum Health and Vitality *77*
Diet *78*

The Rules 79
What to Supplement 81
Probiotics 82
Minerals 82
Killing Pathogens 82
Cleansing Reactions 83

Chapter 11 87
An Alternate Perspective 87
Bioremediation 88
The Environmental Inquiry 90
Ultimate Benefits 91
Living Powerfully 92
Appendix I 95

Appendix II 99
Proprietary Products 99

Appendix III 101
The Body Electrician Clinic 101
Who We Are 101
What We Do 101
How We Do It 102

About the Author 105

Acknowledgements 109

References 110

Testimonials 113
Notes 121

Forward

From the opening page of this marvelous work I was, and still am, impressed by the unique, complete and thorough information provided by Wil Spencer's research over many years of his life. I am also impressed by his personal experience and knowledge of this critical information, which I have found nowhere else; relevant to why the vast majority of the world's population receives such unsatisfactory health care. A sad circumstance which, in my opinion, is a result of conventional medical practices (note the word practices).

It has become abundantly clear to me, that most main stream medical practitioners look at a person with a "medical problem" considering ONLY the observable symptoms, or what results their tests produce. The next step for these, so called, doctors is to consult their "book of symptoms" to see which pharmaceutical "medicine" should be used to treat that symptom or result.

My personal experience with the Veterans Administration doctors has, for many years, been adversarial in their attempts to prescribe cholesterol "treatment" for me, in the form of statin drugs. Every appointment presents the same conversation and each time, I ask the doctor, "do you know, and are you aware, of the dangers and problems with statin drugs?". The predictable answer I have been told time and again (with a somewhat straight face) by the V.A. doctors is; "There is no danger to these, or any other pharmaceutical drugs, or we would not be recommending (actually insisting) that you take them". However, I am well informed enough to know, each and every medical diagnosis has a patent number, ultimately producing financial enhancement.

In addition, for each and every prescription filled, for any patient, the doctor receives a fee from the pharmaceutical company that developed that particular prescription drug, which is also patented.

It is my firm belief that every person on this planet is completely and totally responsible for their own health and health care. Regardless of what main stream medicine would like to have each of us believe, and despite outrageous mal-practice insurance fees, it is the one living in the body who enjoys, or suffers from, the impact of the chosen health care path.

As I read through Wil's book, I am impressed that the knowledge he shares must be divinely inspired. Wil's personal journey was guided through many years of a myriad of physical problems conventional medicine, through the use of pharmaceutical drugs, was not able to cure, but, at best, only relieve some of the symptoms of the many problems he lived with. I understand what Wil has experienced, the problems I have experienced are not dissimilar.

Problems, which, by the way, for me, are being ameliorated with the all natural supplements Wil has suggested. My protocol is based on the results of an electronic hand scan that tests so many aspects of the total and complete health of a person's body. I have had three periodic scans at 2 month intervals which have turned my health and well being completely around.

I can say without hesitation, reservation or lack of confidence, that Wil Spencer and his understanding of what a human body really needs to be healthy, along with the crystal clear and insightful explanations found throughout this incredible book, in my opinion, should be made mandatory teaching and training in all medical and nursing schools. Should that occur, problems with the medical practice would cease, and everyone on this planet could once again be truly healthy. Wil's book should also be mandatory reading in every health class in every high school all around this fragile beautiful blue planet.

The ease of understanding what Wil Spencer has had the courage, experience, knowledge and desire to write about is actually very simple to read and comprehend. Myself, and many others, are waking up from their sleep of generations and learning it is okay to question any person who "practices" medicine, asking, (Wil's sacred question) "why" are you prescribing what you are prescribing and what will it do for me? The information garnered from this book is down to earth logic that supports the realization that good health is a personal mission. It is also our best defense against the risk of living among the many corporate interests affecting life on all levels. People are waking up.

The Divine Creators, in their eternal wisdom, have provided all the natural remedies for any "disease" the corrupt pharmaceutical and chemical companies can invent and then create unnatural, synthetic "medicine" for. The Food and Drug Administration so conveniently approves of the vast majority of drugs, chemicals, etc., which are later proven to be damaging and/or killing the humans, animals and plant life around the world. Could it be that "money" is their prime motivator?

I ask the reader, to just sit back and think about what is really going on around them, besides television, sports, lame-stream media news reports, reality programs, computer games, movies etc. These "pass-times" are in my opinion, just ways to divert real and honest people from understanding and discovering exactly what the bankers, lawyers, doctors and big corporation are doing to the environment and life in general. This planet was given freely to all of us by the Divine Creators for everyone, free of charge.

Wil Spencer's book is an eye opening adventure into what is real and works for anyone who is serious about returning their body to its original healthy state. I have found, through my own personal experience with the supplements Wil has recommended, a tremendous improvement in my body, brain and nervous system.

Neuropathy in my right foot, in my opinion, was caused by the Veterans Administration doctors treating the symptoms of a blocked primary artery, rather than the cause, which they attempted to blame on smoking.

Circulation in my legs has been a problem since 2008. The V.A. Hospital has provided me with surgery to unblock the main artery. When that did not work, there were multiple surgeries placing stents in the arteries. To top it off they replaced the main artery with one from a cadaver.

In 2010 I had a full ultrasound examination for both legs. The ultra sound examination showed that the cadaver artery was still somewhat blocked, although there was enough blood flowing through the artery to make a difference.

The vascular surgeon told me he was going to open my abdomen and see what was causing the problem. Can you imagine the look on his face when I told him quite directly, "There will be no more treating symptoms for me!". I asked him what he thought was the cause for the continued blockage. He attempted to tell me it is because I smoke (only all natural) tobacco.

Having researched that particular subject, before going to the VA vascular surgeons, I then shared with him the results of my research with smokers, some of whom had smoked for over 75 years, of which none have had or experienced the problems I have with blood circulation for the last 5 years.

Once I did the research, I commenced my own regimen of using only natural products, rather than the pharmaceutical drugs they suggested I take. The supplementary arteries had already started to heal and there was still pink splotchy spots on the right foot indicating that some blood was still flowing in and out of that foot, which was verified on the ultrasound results.

It is now May 2013, and the blood flow thru my right leg is normal and the neuropathy caused by the reduced blood flow is healing at a rapid rate since I commenced taking the EMLC supplement two weeks ago. The "pins and needles", foot asleep feeling is gone, and there is only a slight feeling of heaviness now, only when I walk for some time. Wil Spencer's supplements are working and work as they should, healing my body without any experience of side effects.

There are other problems within this 72 year old body that will be healed and I know it will take time. It took a long time for the pharmaceutical drugs and symptom treatment from medical doctors, etc. to cause the damage. It will take some time to heal using the advice contained within Wil's book and continuing hand scans and consultations with him on the results.

There is no question or hesitation on my part, having my experience with Wil Spencer, his advice, supplements and his knowledge, in recommending this book to the world. The information presented in this book can be a life saving investment for any and all who care to take the responsibility for their own health care and quality of life and want the possibility of living to an enjoyable age.

I know that Wil Spencer has the knowledge and experience to stand up for what is right, proper and correct regarding how to repair and keep the human body well. He also has information and supplements that can, and will, correct the Bee Colony Collapse plague. Wil also has supplements for all other kinds of four legged companions that are truly amazing.

I recommend everyone buy two copies of this book, one for themselves and one to give to their medical practitioner, who will probably tell you that Wil has no idea of anything and only their deadly pharmaceutical drugs and supplements can save you.

David L. Corso

Owner and Executive Producer for internet radio station

http://www.wolfspiritradio.com

Prologue

Conspiracy theorists, it's time you were guided out from under your tinfoil hats, out of the darkness, and into the light. No longer relegated to the arenas of theory, of philosophy, of abstraction, **The Underlying Cause of the Unconscious Conspiracy Against Our Health**, will enable you to understand the practical reality of the chemically induced assault on our environment and the assault of that environment on the human specie.

New, serious 'diseases' plague our population today more than ever in its recorded history. The trend can easily be seen that as we as humans moved our civilization from the natural to the industrial, so have we moved deeper & deeper into the dark quicksand of these illnesses.

At first, allopathic medicine, our current "accepted" system, claimed to have all the answers. The answers however were to treat the symptoms through the exceptionally profitable drug model we employ today. As an alternative to this, many have sought more natural treatments through homeopathy or the health food store or alternate health guru, resulting in a HUGE boom in the now industrialized world of alternative treatments.

Dr. Wil Spencer, with all the economic style and common sense practicality of a Mid-Western Farm boy will provide you with the means to step off the merry-go-rounds of these approaches to health. You will be empowered to TAKE CONTROL OF YOUR OWN HEALTH, that health which is UNIQUE TO YOU, and implement practical steps in your daily life to achieve the goal of VIBRANT HEALTH.

The, Underlying Cause of the Unconscious Conspiracy Against Our Health, will give you understanding of the challenges we all face today due to the multifaceted assault on our health, but additionally, it gives a fresh understanding on how to conquer these challenges.

For the impatient or the doubtful, opening straight to, *Chapter 9: Simple Program*, will not only undeniably prove this to you, but will fill the curious mind with a new set of questions and a new set of goals in your mind as you start from a new beginning.

Dr. Wil Spencer will provide you not only with fresh, cutting edge answers, but also with the information necessary for you to find your unique answers in the daily management of your health. This is not a product, or a program, or a doctrine as we currently understand them. This is a commitment to a change of attitude, a change of belief & a change of lifestyle that will unleash unimaginable new potentials in your mind and your body.

Whether you are sick of being sick, or are wanting to get out of the regime of daily pharmaceuticals, or are perhaps just seeking greater health, **The Underlying Cause of the Unconscious Conspiracy Against Our Health**, will provide you not only the knowledge, but the ability to put into daily practice, a nutritional & medicinal strategy to conquer these challenges, and achieve the VIBRANT MIND, VIBRANT HEALTH & VIBRANT LIVES we each deserve!

Robert C. aka - GnoOne

Preface

This work is a compilation of a lifelong voyage and endless hours of research driven by personal pain and illness, followed by clinical experience helping others while looking beyond the seemingly endless list of 2,500 or so symptoms. These symptoms affected not only myself but those closest to me, as well as almost everyone in the industrialized world. The simple question always being: "But why?"

Coming from a typical middle class farming community in the upper Midwest, no one questioned how children were raised. Nor was there much opportunity to see or compare ourselves with the rest of the world. Illnesses were just a part of life that folks accepted and went to the doctor for. There was no concept of heredity or other causative factors for illness; there was just the doc and his fix.

From an early age, along with my own problems, I was surrounded with the sickness and maladies of others. Almost everyone in my community was suffering symptoms of one sort or another. Rashes, overweight, allergies, bowel trouble, acne, and menstrual problems were commonplace. Then there were those with more serious and chronic issues like cancer and diabetes, heart trouble, and depression.

My older half sister went through two kidney transplants and eventual decline to death. My father had an escalating overweight problem with massive sugar addiction cravings, emphysema, and cancer, all the while on an emotional roller coaster. My older half brother and his family were ravaged with health issues including cancer, emotional chaos, joint replacements, and digestive disorders.

My own life was misery; I suffered from 75 allergies along with asthma, chronic bronchitis, and pneumonia. I was overweight as well, had extreme eczema, and my eyesight was failing.

After living with illness for over 20 years, my father's constant empowering words finally began to have an impact. Dad drilled into my head over and over throughout my formative years, "Son, you can do and you can learn anything and everything you want to."

When I reached an age of personal empowerment (thus taking responsibility for myself and my health), it occurred to me that I was in a downward cycle of illness and drug treatment was providing no real benefit. My father's words inspired me to venture out of conventional allopathic healthcare and research alternative, more natural solutions. A close friend, Pat K, first introduced me to wild herbs as a natural remedy.

Learning and understanding practical uses of wild herbs and natural remedies gave me enough relief from those years of illness to set my path to recovery, and to a unique educational experience.

On that path came the opportunity to cross paths, absorb knowledge from, and interact with cultures and people from all corners of the planet, and to form a world view much greater than that of my upper Midwest childhood.

Using herbs and natural remedies brought much greater relief than the doctor's pharmaceutical allopathic medicine. However, these were all remedies for relieving symptoms and had to be continually taken or used or the symptoms would return. This awareness uncovered for me a concept of symptom relief. Relief from symptoms is not a cure and, following reason, there must be an underlying cause for the symptoms to occur in the first place.

My Virgo nature sent me on a quest to discover not only the source of all this disease, but to answer the question, "why". Why society, as a whole, continued toward self destruction, doing things the same ways over and over, somehow expecting different results.

What arose in my awareness was first, the unquestioning trust that people had in the industries themselves. People have a mind set of complete trust in the wonders of doctors and medicines, science, research, and the trend toward convenience and technological improvements. This mindset is so strong, so deep, that it doubts almost nothing said by the doctor, the media, the government, the powers that be. These beliefs are so strong that people have learned to be satisfied with answers based in mere symptom relief.

On the other, darker side of this implicit trust I observed self-serving, commodity based corporate entities employing technologies that literally create symptoms and instill fear into an already over stressed public.

Looking for the actual root causes of my own health challenges, and later working with others in my clinic, revealed that in almost every case there is an underlying cause that can be traced to either a GI tract or an electrical imbalance. In *Chapter 10: Dietary Parameters of Workability*, you will find what has developed as a very simple and easy understanding: how our body's ability to maintain our life force, stay healthy and vibrant, and nourish our animation comes from our diet and our connection with nature.

Gone are the days when the whole family participated in raising, handling, processing, preparing, and being part of, the whole food chain. Sadly missed are the simple pleasures of living with the natural balance of the rhythms of nature. While it is probably unreasonable in our modern lifestyle for more than the few native and grass roots cultures (those who still hold fast to what works in tandem with nature in the face of industrialized convenience), to actually live this natural life, it is possible to manage good health in an industrialized world.

For most people in the industrialized world, food has come to be something we simply eat when we are hungry. The food chain has become industrialized in order to serve the demand for convenience, convenience at the cost of real nourishment, convenience that misses vital and critical elements for health.

One of the elements of my unique educational process has been the serendipitous events of my life, which gave me firsthand experience of the massive changes in the food industry. Life took me from a simple farm family through the onslaught of the industrial takeover of small farming and the transition to careless, greedy, artificial mass production of plants and animals.

Transporting crops and animals during these changes afforded me the ability to see behind the scenes in the food factories and in the shipping methods. These experiences provided invaluable firsthand experience, confirming (and in some cases exposing) many both unconscious and intentional practices that severely affect the quality of food and health.

Searching this new food chain for the missing ingredients that once kept us healthy, we find a huge deficit in a broad spectrum of bio-available minerals and microorganisms. Minerals are necessary building blocks in which all of the electrical frequencies needed for vibrant health are carried. Microorganisms are needed to synthesize minerals, vitamins, and amino acids from our food to our bodies, while protecting and enhancing our immunity.

The new science and its new understandings have served to clarify and verify my views of the body and health. Dr Dawson Church, PhD author of **Genie in Your Genes**, explains the components of the body's electrical system as comprising the DNA in our genes, the rod and cone cells at the back of our eyes, the myelin sheath of our nerve cells, the collagen molecules that make up all our connective tissues, our muscle tissues, and the phospholipids of all cell membranes. All of these electrical components are built and maintained through the ingestion of mostly animal proteins.

In gratitude for my life's path, which has taken me from personal illness, to vibrant health, to the personal empowerment that comes from making a difference for others, I humbly share what I have learned and come to know as truth: the current trends in food and health are not working and, in fact, are failing at an escalating rate.

Furthermore, life will continue on its current path of decline until these deficits are recognized and corrected. The concepts and information that are being shared in this material can change that.

It is my intention and hope that you will come to understand for yourself that your good health is a matter of your own choices. Your health should not be left to practitioners and institutions who misguide with symptom relief only, whose primary concern is their own corporate expansion and development. The information herein exposes (enough for you to see) the imbalances created by depleted soils, toxic environments, and processed, empty foods. It exposes the rampant disinformation from our supposed experts in these fields.

Ultimately, the answers I have found and offer here for your discovery point to a systemic condition in all of life. Our return to vibrant living requires the same approach to the same imbalance that is caused by the same degradation that is in all living systems on the planet: as above, so below and as within, so without. In order to succeed in returning health to our ill environment, we only have to look within.

E. Wil Spencer

Chapter 1

The Unconscious Conspiracy

Volumes have been written (and referenced here on page 110) documenting and explaining the decline and degradation of our environment, specifically the air, water, soil, plants and animals. Pollution is a well known fact and we have reached a point where it has undeniably caused a systemic health crisis among the people on this beautiful planet. Attempts at resolving the symptoms and diseases that continue to arise have consistently been taken from an allopathic approach of symptom relief only. This approach is fast proving itself to be not only ineffective, but even a cause of further decline. It would now seem valuable to look at another perspective in order to piece together the puzzle of our unconscious demise. Then we can see the simplicity found in the root of this matter of our health.

The unconscious conspiracy of which I speak is really the industrial road to progress. This is a road which has evolved into a massive expanse of speeding superhighways spreading into every facet of human existence. The trip has taken us thousands of miles, moving so fast we haven't even noticed the scenery, much less the effect our travels have had upon it. We ponder issues of pollution and health, unable to see the connections whizzing past our narrow window view.

We have been caught up in looking at the complexities of the many issues we are dealing with as a species: surplus war chemicals, industrialization, big business, big agriculture, big "pharma", etc. Following the money trail of progress and development has left us a legacy that is mind-boggling, nearly impossible to understand. However, real solutions, when finally found, are always simple, and work at the source of the trouble.

Finding the Simple Solution

Finding the simple solution to the health crisis we face requires an answer that not only has curative effect, but also provides lasting prevention for humans as well as the air, water, soil, plants, and animals that constitute the food chain.

Looking at the thousands of symptoms of illness we find in humanity today, along with degradation of all kinds, is gross complexity. Yet the body is a creation that was placed in a much simpler environment than we have today, an environment in which the body had the ability, and was given the tools in nature, to maintain its own health, immunity, and wellness. So, rather than looking so intently at what has been added and done to cause all our problems, perhaps the simple solution lies in finding what is now missing, the absence of which has allowed such pestilence to occur. Pestilence is in the human body, evident in these thousands of symptoms causing suffering and disease.

Considering some parallels in the natural world, we can learn much from the honeybees, which are plagued with pestilence of their own. Beekeepers have come to a crisis losing a third or more of their hives annually. The bees are suffering mites, nosema, foul brood, bacterial and fungal outbreaks, Colony Collapse Disorder, etc. Proof is in, that synthetic agricultural pesticides (from genetically modified crops that contain systemic pesticides like neonicatinoids) are causing neurological disorders and confusion, disrupting the bees' immunity and function.

This is one of many similar scenarios in which the hives become toxic and the bees abandon their homes. All the issues affecting the bees are indicating that nature is out of balance. This weakened and out of balanced system is showing us with the appearance of mites, bacterial and fungal outbreaks, Colony Collapse Disorder, etc., that it is regaining its own balance by eliminating the sick, unhealthy, toxic, and weak.

Elsewhere, we have cattle eating an unnatural diet of grain, including genetically modified corn, cottonseed, and soy, as well as other industrial waste. Their manure is used as fertilizer, causing E.Coli and other outbreaks in the environment. We have meat tainted with E.Coli, Salmonella, Listeria, and Clostridium causing mass food recalls and human illness. These are pathogenic mutations of naturally occurring bacteria gone extremely out of nature's balance.

In nature's effort to recycle a weakened system, in this case the cow's gut, the bacteria overpopulate in order to decompose and recycle a sick, dying or dead organism. We also see proof that those same cattle, if fed a natural diet of healthy grasses from a pasture for just two weeks, would regain their natural balance and no longer pass infectious manure or unhealthy protein food products into the food chain.

For over 80 years, industrial surplus and waste have been applied to crops in the form of NPK (nitrogen, phosphorous, and potassium). We see these plants lacking nutrition and weakened to the point where they cannot resist infestation from pestilence such as corn bores, boll weevils, potato bugs, and other harmful insects. The conventional fix for the pestilence is symptom relief in the form of another chemical pesticide... and the cycle continues. The point to note here is that this is another scenario of nature out of balance, calling in the recycling troops, which show up as the pestilence.

What we can learn from looking this way into a variety of areas of nature at work are the parallels running through all life forms. Each life form seems to have its own form of modern pestilence that, when looked at more closely, shows common signs of an underlying cause: nature is performing her natural recycling clean up of weakened, out of balance, or you could say "sick systems."

Natural Laws

Nature functions without judgment according to its own laws. Life forms grow, mature, reproduce, degrade, and recycle. Nature constantly renews the biological material by using one stronger biological system to break down another weakened or sick system and use it for food to fortify its own strength.

If we can look at what is happening with people's health from this aspect of nature, we can identify nature's recycling methods at work on us. From this perspective we can see that we are in a spiraling cycle: we are attempting to resist the natural laws, but we are resisting with symptom relief, not with cures. Although we may succeed at doing it this way for a very long time we can be assured we will not be well in the process.

As I said earlier, the human was created for wellness, and was put into a natural environment that had all the physical makeup and tools necessary for that wellness. We have made changes to the natural environment that have caused nature to lose alignment with our wellness. In other words, we have become out of balance with nature. Our food chain is corrupt, unhealthy, and out of balance.

Organics have been, so far, our best effort to overcome degradation of food. As such, organics have been effective at resolving the problem of chemical effects and content in the food supply. However, we see relatively small increases in actual nutritional value, indicating that removing what has been added is not recovering what is now missing. In other words, non-use of synthetic chemicals does not sufficiently increase available nutrients in food.

Life Originates in the Soil

Since all life originates from the soils, all of life is really a mirror of the overall condition of the soil. In the 1920s a census was performed on all soils of the industrialized world, which concluded that soil then contained 20 to 40 inert available minerals. A similar study done in the late 1990s, showed a significant decrease to only 18 to 25 available minerals present. In his book, **The Root of All Disease**, Elmer Heinrich quotes Gary Price Todd as saying that for optimum health a minimum of 60 minerals is required for humans and 45 for other mammals. He also states, "Sick soil causes sick plants which cause sick animals and ultimately sick human beings."

Those using synthetic pesticides and industrial waste for

treatment of soil fertility have overlooked the paramount factor of how those additives affect the availability of life in the soil. Organic farming, though recognizing the detriment of pesticides and industrial waste, is also overlooking this paramount issue of life in the soil.

The life force no longer flows in a smooth stream of energy, from the sunlight to the soil, through the plants and animals, and on into the animating nourishment that makes a human being thrive. In our quest to outmaneuver Nature through our own improvements, we have isolated ourselves from our natural Earth connection, and in so doing have disrupted the balance she provides. Our environments, both those within our bodies, and those in which we live, are corrupted and severely out of balance.

We face a time in which, for the first time in our own evolution, it is predicted that our children will no longer be healthier than their parents. In addition, they will not enjoy the increased longevity of previous generations. This will be the case unless we can recognize what has occurred among us and take the necessary steps to return our essential nature.

Sun, water, minerals, and microorganisms give life in the soil. Microorganisms give life to life. In the soil, microorganisms break down the minerals that are so essential to life, in an enzymatic action that makes them available to the plant. In the process of photosynthesis, the life giving minerals become nourishment for the plants. The plants then produce a form available to mammals, which the mammals can absorb with the help of microorganisms in the gut. Microorganisms in our gut keep us clean through natural chelation of toxins and provide 80% of our immunity. They synthesize all needed vitamins and perform electrical recognition of the building blocks of amino acids and minerals for the body to use as it sees fit.

A Healthy GI Tract

Although the human body is a complex machine, we can understand the basic requirements we need to give our attention to by examining a little more closely the digestive tract and the processes that take place therein.

Nourishment begins with the thought and with the first sensing of food. Digestive solutions prepare for what we are about to eat before a bite is taken. Gastric juices are first added in the mouth. Chewing is the process of breaking down the food into manageable-sized pieces that the stomach can handle. After swallowing the well chewed food and gastric juices, the stomach, microorganisms, and enzymes break open the cellular structures and ferment (or rot) the stomach contents in preparation for moving into the small intestine. This task, to be optimal, must be accomplished in the two to four hours that the food remains in the stomach. Otherwise, unprepared material is passed through to the small intestine, which then has to alter its normal function to accommodate the error.

While in the stomach the food is being broken down into the parameters of microscopic activity. At this stage something very magical and astounding happens. Imagine cracking a raw egg and a lightning bolt is released in the instant the shell breaks through. Breaking open the cellular structure of food closely resembles a lightning bolt escaping. The magic is in understanding that this is how the body gets its animation, its electricity, or its batteries charged. Every molecule of (electrically bio-available) food has a spinning lightning bolt, or a spark, within it. The spark is released from the molecule and magnetically drawn into the tissues of the stomach lining to become the body's living force.

The lining of the small intestine is home for microorganisms and where enzymes are created and nutrients such as minerals and amino acids are released into the body through the lining. Should the small intestine receive the chyme (well chewed and fermented food and gastric juices from the stomach) according to design, a multi function process takes place. Vitamins are synthesized by the microorganisms and supplied to the body. In addition, the natural negative charge of the plant sourced minerals passing into the body magnetically attracts the positively charged heavy metals and other toxins (like pesticides and those found in cosmetics) that are pulled from the blood and elsewhere throughout the body.

This negative charge pulls the toxins and heavy metals through the intestinal wall lining. They are then consumed by microorganisms which are also negatively charged, attracting free radicals, toxins, and heavy metals (including pharmaceutical drugs, pesticides, industrial food additives, and the like). This is a natural chelation process that takes place in a healthy and balanced body.

As the chyme passes through the rest of the intestines, excess moisture is drawn out to produce a healthy consistency of fecal matter. Microorganisms are present in this entire eliminating phase to prevent putrefaction or any other pathogenic reaction.

The healthy digestive tract is a perfect system for nourishment, immunity, and healthy elimination, provided that what is put into it is aligned with what the body needs, and that there is sufficient population of the broad spectrum of microorganisms available for proper processing.

Basic Ingredients for Health

The body literally gets everything it needs to maintain excellent health from just a few basic ingredients. These are the amino acids, a broad spectrum of plant sourced minerals and soluble fiber, and a broad spectrum of microorganisms present in the gut.

Ideally, amino acids are provided from mainly animal protein and fat. Minerals are obtained from fruits, vegetables, and animals (that is, if the animal has been nourished with plants and with soluble fiber coming also coming from a vegetative source). The human body does not require carbohydrates at all, a subject addressed more fully in *Chapter 2: Grains, Nuts, Seeds, and Tubers*.

Most people today are lacking the majority of the microorganisms vital to intestinal health. This is because these microorganisms are no longer in the soils, thus they are no longer in the food chain. Minerals are depleted from the soils, and there are now too few microorganisms to synthesize them into the plants. Therefore, neither minerals nor microorganisms are found in the fruits or vegetables. Neither are they found in the animals, as the animals get their minerals and microorganisms from the plants, as well as from contact with the soil.

Amino acids that come from conventional meat sources now come from animals that are eating unnatural diets, so the meats enter the food chain with a lot of other concerns and deficiencies as a result of the animals' diet and environment. This is exacerbated by the unnatural and unhealthy handling of animal meats and other animal products in the conventional processing methods.

Even an organic food chain in today's world cannot provide an ideal or acceptable diet because of the many deficiencies now present in the soil. Thus, malnutrition is at an all-time high. This is the basis for the vast majority of the health issues experienced worldwide today. Thus, the simple solution to this global health crisis is as straightforward as eating correctly.

Hippocrates said that, "A wise man should consider that health is the greatest of human blessings and learn how by his own thoughts to derive benefit from his illness." He also said, "If we could give every individual the right amount of nourishment and exercise, not too little and not too much, we would have found the safest way to health," and, "Everyone has a doctor in him or her; we just have to help it in its work. The natural healing force within each one of us is the greatest force in getting well. Our food should be our medicine. Our medicine should be our food." When Hippocrates lived, from 460 BC to 377 BC, there were not the conditions of industrialization, chemicals, genetic manipulation, vital nutrient depletion, pollution, conventional agricultural practices, etc. that we see today. Nonetheless, he did have the simple wisdom concerning true health and vitality to which we need to return.

Eating correctly today will require supplementation of a

broad spectrum of minerals and a broad spectrum of micro-organisms, combined with a diet of non-GMO (genetically modified), preferably organic fruits, vegetables, and animal products as outlined in *Chapter 10: Dietary Parameters of Workability*.

Eat for life, and then give life time to heal and restore. We are created as incredible, miraculous, vital wellness machines that are able to live and function through long suffering of disease, illness, and depletion. But we don't have to! As long as there is life in us, there is the ability to regain our heritage of vibrancy, energy, vitality, health, and love.

Chapter 2

Grains, Nuts, Seeds, and Tubers: An Electrical and Biological Nightmare

This chapter briefly explains concepts and clinical experiences that will help you understand the benefit of refraining from ingesting grains, nuts, seeds, and even beans, and potatoes or tubers of all kinds.

From an elemental view, grains, nuts, seeds, and tubers contain a wide range of nutrients and fiber. The issue with the consumption of grains, nuts, seeds, and tubers lies in their biological and electrical incompatibility with the human body. Some insight into the world of botany will help you understand the difference between the elemental facts and natural biological truths.

Anti-Nutrients

Most of the nutrients and fiber in grains, nuts, seeds, and tubers are indigestible, and when eaten, are processed through the human system as anti-nutrients. Anti-nutrients are chemical combinations of actual nutritional elements. These chemical combinations are presented in forms that block our bodies from obtaining the nutrition the anti-nutrients contain. Instead of recognizing them as food, the body perceives and processes them as what they are, which is toxin.

Plants and animals reproduce. Animals have the advantage of movement to keep away from predators. Plants, on the other hand, do not have the ability to move and escape from being eaten. Instead, nature has provided for their self-preservation through the evolution of several innate survival capabilities. One of these is the production of chemicals referred to as anti-nutrients, which are natural pesticides stored in the seed, their offspring, as a way of survival. These natural pesticides cause severe health degeneration and illness to all those creatures that eat the seeds, the offspring of the plant. A causative factor in the escalating rise of infertility, digestion, and immunity issues in our population today is the ingestion of these natural pesticides.

These anti-nutrient pesticides are also enzyme blockers, which cause the predator to eventually lose the ability to create enzymes for digestion. This loss can ultimately cause the predator's death because it is no longer able to break down and use its food. The enzymes the body stops making are protease, pepsin, amylase, trypsin, lactase, chymotrypsin, and lipase. They are also the main enzymes needed by an omnivore or carnivore for survival and proper digestion. Can you see the negative impact on the human, who is an omnivore, with the ingestion of grain, nuts, seeds, and tubers?

Lacking these vital enzymes causes putrefaction of meat, or animal protein, in the gut as the body attempts to dispose of undigested waste the only way it can. The microbes need to break down what has not been fermented in the natural enzymatic process and cannot be digested. The only solution available is rot, decomposing the protein into putrescine and cadavarine, the two sickeningly aromatic smells of death. This leads to impaction of the small intestines.

Considering that the whole digestive tract from mouth to anus is an extension of the skin and actually part of the exterior surface of the body as one continuous loop, it makes sense that putrefaction in the gut can relate to irritation and allergic reaction on the skin. A vulture, who feasts on such rotten materials in a scavenged carcass, is protected by nature with a bald head and an intense microbial balance.

The function of these enzymes is to break down food in the human GI tract. Without them we are unable to receive the life sustaining amino acids and minerals that comprise the animal fats and proteins. When these amino acids and minerals are no longer available to us, our bodies will slowly degrade, become ill, and ultimately die of malnutrition.

High Starch and Insoluble Fiber

Today, everywhere, there is evidence of malnutrition and digestion disorders. Some of the named diseases are gastritis, allergies, Attention Deficit Disorder (ADD), dental cavities, constipation, hemorrhoids, Irritable Bowel Syndrome (IBS), ulcers, GERD (Gastro Esophageal Reflux Disease), iron deficiency anemia, ulcerative colitis, Crohn's disease, eczema, leaky gut syndrome, acidosis, acne, colon cancer, osteoporosis, arthritis and candidiasis to name a few.

High starch content is another problem when consuming grains, nuts, seeds, and tubers. In order for humans to use starch, it first has to be converted to glucose or blood sugar. This conversion from starch to glucose requires the chemical/hormone insulin, which is created in the pancreas.

In a diet high in grains, nuts, seeds, and tubers, we see a parallel high insulin level in the body. High insulin levels have many negative health effects on the human body, including diabetes, acne, chronic infections, celiac disease, most autoimmune disorders, anemia, dehydration, human growth factors, seborrhea, impotence and premature ejaculation, prostatitis, high blood pressure and other heart diseases, dementia, nerve damage, obesity, hypoglycemia, Graves' disease and other thyroid failures, fatty liver, depression, anxiety, hyperinsulinemia, malnutrition, gall bladder diseases, hormone level disruption, insomnia, migraine headaches, blurred vision, chronic fatigue syndrome, kidney diseases, dizziness, sugar cravings, muscles weakness, fainting and even coma.

Additionally, grains, nuts, seeds, and tubers contain insoluble fiber. The definition of insoluble is an element that cannot be dissolved; not soluble. The human body is only able to use elements that are soluble. In an attempt to ferment or digest insoluble fiber, huge amounts of acids are produced in the stomach to break down the fiber into a usable element. The process of fermenting or digesting this insoluble fiber is very time consuming and cannot be completed before the fiber is moved out of the stomach. The fermentation or digestion process then must carry over into the intestines, creating additional health problems. Problems that can result from undigested insoluble fiber are dysbacteriosis, ulcers, ulcerative colitis, IBS, leaky gut syndrome, GERD, atherosclerosis, hemorrhoidal disease and anal fissures, cancer, H. Pylori overgrowth, appendicitis, malnutrition, bloating and gas, constipation and/or diarrhea, enteritis, hernia, duodenal epithelium, dental cavities, esophagitis, Barrett's disease, halitosis and gingivitis.

The overwhelming amount of sugar, insoluble fiber and starch working its way through the GI tract is food for opportunistic bacterium, fungi, viruses, and yeasts. These bacterium, fungi, viruses and yeasts naturally occur in our GI tract and are a very integral part of our immune and digestion systems. However, when allowed to overeat and proliferate on sugars, insoluble fiber and starches these bacterium, fungi, viruses, and yeasts will become pathogenic and cause a very unhealthy internal environment. In this feast of optimal food for them, E.Coli, Salmonella, Candida, Listeria, Streptococci, Klebsiella, Clostridium and Aspergillus will eventually cause the GI tract, and thus the whole human, to suffer greatly.

There is no such thing as a necessary carbohydrate. This may be a challenging statement to accept, yet I repeat: there is no such thing as a necessary carbohydrate. Drs. Eades wrote in their book, **Protein Power**, "The actual amount of carbohydrates required by humans for health is zero." The doctors eloquently dismiss the carbohydrate myth along with several other dietary myths in satisfactory detail in their book. The valid point to their discussion is that every cell in the human body has the ability to create its own sugar or energy as it requires. Thus the ingestion of carbohydrates, especially in the quantities they are generally consumed today, serves only to damage the internal environment of the human body.

While they can be substantial sources of carbohydrate, fruits, and vegetables, are better understood as dietary components in another discussion devoted to the class of food they occupy.

Finally, there is the lectin connection. In her book, **The**

Vegetarian Myth, Lierre Keith explains very well the point of lectins, another anti-nutritive factor in seeds that interferes with protein digestion. In a diet high in grains, nuts, seeds, and tubers, the lectin intake is also very high. Lectins do not break down in the stomach; they actually attach themselves to the intestinal walls, causing inflammation and irritation. This happens as the presence of high amounts of lectins on the intestinal membrane damages and shortens the villi, upsetting the balance of intestinal microorganisms and accelerating cell death. The overwhelming number of lectins in the GI tract then has the ability to transfer through the damaged mucosa lining and infect the rest of the body via the blood and lymphatic systems. Once lectins are in the blood and lymph the devastation is massive, leading to autoimmune disorders like rheumatoid arthritis, multiple sclerosis, thyroid inflammation, Crohn's disease, food and chemical allergies and sensitivities, psoriasis, type 1 diabetes, mellitus, lupus, skin rashes, celiac disease, and asthma.

Hopefully you now see the importance for humans not to consume grains, nuts, seeds, and tubers in the majority, if at all, in our diets. The overall devastation to the human body is unmistakable and undeniable. The concept, of not eating grains, nuts, seeds, and tubers may be challenging to your belief structure, but the effects on the human body are real and clear, as we have seen, working with thousands of clients in our clinic.

Historical Truth

This concept of eating little or no grains, seeds, nuts, and tubers becomes even clearer when we look at ancient

Egyptian history. There is a misconception derived from viewing the hieroglyphs of beautiful figures represented as god-like entities, that this is the way the Egyptians actually appeared, or at least the population of the hierarchy. The common Egyptians were predominately farmers, and all Egyptians consumed mostly grain, nuts, seeds, and tubers. Their staple foods consisted of coarse ground whole grain flat breads along with vegetables and some fruit. The Egyptians consumed fish and poultry in very small quantities and they ate virtually no red meat. They did use goat and sheep milk to make cheeses, however, milk was a beverage reserved for the people of higher social rank, as owning these animals was a privilege of wealth and status.

Reviewing the archeological and paleontological records of the Egyptian mummies reveals that they had health issues similar to what we have today. They suffered dental cavities and gum disease. Their teeth were weak and deformed from the lack of adequate protein intake and showed severe deterioration and wear. Their teeth were ground flat and worn off from chewing huge amounts of insoluble fiber, which takes a large amount of chewing to break down so the stomach can do its job. This also indicates high acid levels in the body, which eat away at the enamel making the teeth weak.

Another sign of inadequate protein intake noted was in bone density and size. Bones, relatively easy to study, hold a lot of information as to the overall size and health of a population. The Egyptians' overall size diminished over a 3,000 year period of time, their bones becoming smaller and weaker.

Records from the paleopathologists suggest other common diseases between the Egyptians and folks today such as obesity, high blood pressure and other heart problems, stroke, and diabetes.

An anthropologist from the Smithsonian Institution, Dr. Kathleen Gordon, writes in one of her papers, "Not only was the agricultural 'revolution' not really so revolutionary at its inception, it has also come to represent something of a nutritional 'devolution' for much of mankind."

In her book, **The Vegetarian Myth**, Lierre Keith quotes Dr. Loren Cordain's article, "Cereal, Grains: Humanity's Double Edge Sword":

"Cereal grains as a staple food are a relatively recent addition to the human diet and represent a dramatic departure from those foods to which we are genetically adapted. Discordance between humanity's genetically determined dietary needs and his [sic] present day diet is responsible for many of the degenerative diseases which plague industrial man.... [T]here is a significant body of evidence which suggests that cereal grains are less than optimal foods for humans and that the human genetic makeup and physiology may not be fully adapted to high levels of cereal grain consumption."

Chapter 3

Lactose Intolerance

A disease of the 20th Century, lactose intolerance, is a product of high tech pharmaceutical think tanks, whose sole function is to create medical names for symptoms manifesting from the successive degeneration of our health, which is due to the underlying causes addressed throughout this book. Once named, a disease can then receive a proper medical protocol of symptoms, diagnosis, and treatment.

The symptoms of lactose intolerance present similarly to those of flu, IBS, and other "allergies." These symptoms can be nausea, abdominal cramps, diarrhea, bloating, gas and flatulence. However, in the cases of lactose intolerance they occur in conjunction with diary consumption.

In a healthy human gut, the enzyme lactase is produced in the lining of the small intestine through a symbiotic relationship between the broad spectrum of microorganisms, minerals, and available amino acids. Lactase then converts lactose—milk sugar—into usable nourishment and energy for the body. When the ability to manufacture lactase is deficient the lactose is improperly fermented causing symptoms of illness.

The ability to manufacture lactase is lost in a diet high in grains, nuts, seeds, and tubers (see *Chapter 2: Grains, Nuts, Seeds, and Tubers*). The natural pesticides in the grains, nuts, seeds, and tubers, or anti-nutrients, cause a reaction whereby the body loses the memory of how to create needed enzymes, in this case, lactase.

As the body is then unable to healthfully process dairy products, so too does it then lose the benefit gained from consuming them. Sufficient intake of calcium, phosphorus, riboflavin, and vitamins (A, D, K, and E) from bio-available sources becomes a serious concern. Weston Price DDS, Denie Hiestand (author of Electrical Nutrition) Saul Liss PhD, Drs. Eades (authors of Protein Power) are among those who clearly state that these nutrients, especially vitamins A, D, K, and E are only bio-available to humans through consumption of food from animal origins. Dairy sources, whether cow, goat, sheep, or any other mammal, (even human raw milk products), are the best and easiest sources of these necessary nutrients.

The Mayo Clinic publication, "Lactose Intolerance" (http://www.mayoclinic.org/diseases-conditions/lactose-intolerance/basics/definition/con-20027906 - 1998-2016) identifies three types of lactose intolerance:

Primary lactose intolerance: Normally, your body produces large amounts of lactase at birth and during early childhood, when milk is the primary source of nutrition. Usually, lactase production decreases as the diet becomes more varied and less reliant on milk. This gradual decline may cause symptoms of lactose intolerance.

Secondary lactose intolerance: This form of lactose intolerance occurs when the small intestine decreases lactase production after an illness, surgery, or injury to the small intestine. It can occur as a result of intestinal diseases, such as celiac disease, gastroenteritis, or an inflammatory bowel disease like Crohn's disease. This type of lactose intolerance may last only a few weeks and be completely reversible. However, if it's caused by a long-term illness, it may be permanent.

Congenital lactose intolerance: It is possible for babies to be born with lactose intolerance. This rare disorder is passed from generation to generation in a pattern of inheritance called autosomal recessive. This means that both the mother and the father must pass on the defective form of the gene for a child to be affected. Infants with congenital lactose intolerance are intolerant of the lactose in their mother's breast milk, and have diarrhea from birth. These babies require lactose-free infant formulas.

Additionally, the Mayo Clinic's prognosis is, "You can control symptoms of lactose intolerance through a carefully chosen diet that limits lactose without cutting out calcium, and possibly by taking supplements."

Symptom Relief Is Not Cure

An imbalance severe enough to cause lactose intolerance will sooner or later develop other symptoms of a nature related to the root cause of the imbalance as conditions progress.

Reversing lactose intolerance naturally is relatively simple, requiring only the willingness to make necessary changes and implement proper supplementation. Since consumption of grains, nuts, seeds, and tubers interferes with enzyme production, they need to be eliminated from the diet and enzymes need to be supplemented until the body eventually regains the ability to "remember" the process to synthesize its own. Also, supplementation of a broad spectrum of microorganisms, minerals, and additional amino acids must be present for synthesis of lactase.

The enzymes I recommend are specially formulated to work in the most effective way with the human body. These enzymes have been formulated with the biological and electrical availability in mind.

Chapter 4

The Candida Story

Candida, a single celled amoeba type fungus, causes one of the most misunderstood and probably the most proliferate and widespread multi symptomatic problems affecting vibrant health in the human body today.

First brought to light in 1915, when Candida was found to exist in the human body. Specifically, it was identified as one of the naturally occurring colonies of microorganisms found in its natural, healthy parameters in the human gut. Candida remained then a non-invasive resident, simply noted as existing, with samples stored in formaldehyde, until long after the 1940s which is when its actual metamorphosis took place.

In 1953 the Japanese Medical Authorities notified our American Medical Authorities that they were documenting an explosion of seemingly unconnected multi symptom diseases and maladies in the populations surrounding Nagasaki and Hiroshima. The only common denominator among these conditions was notably elevated levels of Candida.

American authorities called this a coincidence, and dismissed any connection between the population's growing epidemic and Candida.

In the late 1950s, the Japanese returned with an undeniable documentation of more than 5,000 case histories which were presented to the Mayo Clinic in upstate New York.

The electron microscope had been invented by this time, and American scientists were unraveling the secrets of genetic codes. Being undeniably forced to revisit the Candida issue, the stored 1915 Candida samples were genetically compared to the then current samples. Evidence was clearly established that the current, post-atom-bomb samples of the simple one-celled fungus had been mutated by low-level radiation.

Atmospheric nuclear testing came to a halt with the knowledge that low-level radiation has the ability to mutate and change forever the very basis of our biologic terrain: our very ability to be alive.

Biological Disaster

The next link in this chain of events, which caused the largest biological disaster since World War II, was the discovery, and, in short order, widespread use of antibiotics. It did not take long for the medical world to realize that the use of their miracle drugs, antibiotics, wreaked devastation upon intestinal microorganisms that are the basis of human health and immunity. However, by the time they did realize this sad fact, it was impossible to publicly announce the existence of complications with their fair-haired money making antibiotics. Pharmaceutical business could not admit that its biggest profit maker to date was creating more disease than it cured.

As the use of antibiotics became widespread, industrialization was also growing and expanding its use of post war surplus chemicals in agricultural applications. So, at the same time the human biological terrain of balanced microorganism protection was killed off with antibiotics, the widespread destruction of microorganisms in the soils and crops affecting animals as well as humans, was taking place.

The loss of our broad spectrum of microorganisms, (many of which are transient and must be continually supplied by nature through direct contact with soil, dust, air, etc. as well as through the food chain), left the mutated Candida in the intestinal tract unchecked, and able to outgrow its natural healthy parameters in the gut.

Candida has taken advantage of its opportunity to increase in population, becoming a systemic problem invading first the body fluids and connective tissues, and then mutating even further in this new environment. The new mutation is able to form colonies throughout the body, living in and on organs, mucous membranes, skin, and any damp, moist area such as the genitals. It invades the lymph, lungs, and upper respiratory systems, and lives on and around cancers and tumors.

All told, we now have some 2,500 symptoms that can manifest with the presence of Candida, all of which are present in our society today. Although too many to list in entirety, it suffices to say that things like athlete's foot, excessive phlegm, pancreatic stress, heightened thyroid problems, elevated acid levels, and the compulsive urge to eat grains and sugars (which leads to obesity and another accompanying list of complications), are precursors to generational disorders, syndromes and maladies such as Attention Deficit Disorder, chronic fatigue syndrome, fibromyalgia, diabetes, cancer and arthritis.

By the 1960s there was significant recognition in the

natural health industry of the seriousness of the Candida issue as an intestinal microbiology problem. In a desperate attempt to offer what relief they could for people suffering Candida symptoms, the natural therapists routinely recommended Acidophilus. Acidophilus was the only bacteria recognized as part of the human intestinal terrain that was commercially available. Acidophilus and its close relatives, L. Planetarium and L. Salivarius, do help rebuild a healthy intestinal flora. However, they are but a tiny fraction of a healthy spectrum, and can do nothing for Candida existing outside the intestinal tract.

Bacillus Laterosporus

In 1982, an agricultural scientist named Boyd O'Donnell was on sabbatical in Iceland. He noticed that the native crops of corn were 12 feet tall and had at least three corncobs on each stalk. On another field, there were tomatoes so large he could not get his two hands around them. Upon inquiring with the local Indians as to what their special seeds were, he found that they used the same common farming seeds as the North American farmers, and it was explained to him that the winter had been mild and the summer had arrived early. This caused the ice to melt back and expose more of the tundra than usual. This freshly exposed soil, which had not been exposed since before the last ice age, was possibly the only virgin soil left on earth and in it they had grown these amazing crops without the aid of any chemicals or fertilizers.

O'Donnell took a soil sample back to California. The only thing he found different in his analysis of the soil was an, as yet, unidentified soil-borne bacterium that also populated the plant samples he brought back with him. He spent the next 15 years learning about this bacterium (nearly going bankrupt twice, trying to fund his research) and attempting to figure out production methods for his newly discovered Bacillus Laterosporus (BOD strain).

Used in the 1980s as an agricultural spray, the BOD strain made the crops healthy and vigorous; the farmers loved it! But delivery was a problem; once the crops were processed and packaged, the bacterium was short lived or dead on arrival. Although it has proven to be one of the most difficult, finicky bacterium ever known, Boyd persisted and identified its uniqueness, finally finding a way to cultivate and freeze-dry this bacteria safely so as not to lose its advantageous qualities.

His persistence paid off, and he was granted a worldwide license and patent on this natural bacterium, one of only four natural bacteria to be honored in this way.

One of the chance discoveries made by Boyd's team of researchers was that Candida had no resistance against his newly discovered strain of bacteria. Human trials were carried out, and in each case systemic Candida loadings of the connective tissues and body fluids dropped dramatically when the patient took the BOD strain of bacillus.

As a result of the initial trials, two major New York labs were commissioned to do extensive double-blinded trials. Results were that the BOD strain was found to be the only known natural bacteria to lower Candida levels in the human body.

Despite these findings, the BOD strain remains virtually

unknown or recognized in the alternate health care industry, and is completely ignored by modern allopathic medicine.

Today, drugs like Diflucan, Fluconazole, Nystatin, and other synthetic compound anti-fungals are the medical industry's best offering for Candida relief, even though long term effects of these compounds stress the liver and the intestinal imbalance becomes far more extreme. So, although they are mildly effective on Candida and other fungal overgrowths initially, they also set up the conditions for a renewed overgrowth once the prescription is completed.

The natural industry is offering similar symptom relief with grape seed extract, pycnogenol, colloidal silver, tea tree oil, caprylic acid, olive leaf extract, coconut oil, garlic, pau de'arco bark, oil of oregano, and Oregon grape. These remedies, though not harmful to the body, can only moderately relieve symptoms, doing nothing for the actual systemic problem of a Candida overgrowth.

Chapter 5

Candidiasis

Candidiasis is the medical term used to describe a yeast overgrowth. Other terms for this condition are Candida Albicans, Candida, or sometimes it is just termed yeast overgrowth, or even just yeast. This is by no means a new medical problem. In fact, it has actually been around for centuries. However, it has become a chronic, modern-day medical dilemma that seems to be increasing rapidly.

Yeast is an integral part of life, as it is present in food, it is found on a majority of exposed surfaces, and is in the very air that we breathe. Yeast, along with fungi and molds are nature's recycling agents. Wherever there is organic matter that is dead, dying or weakened these recycling agents manifest and grow to complete the cycle of life by breaking down, decaying, and ingesting the weakened material.

There are approximately 250 types of yeast, many of which are parasitic in the human body. The major yeast strain present in the human body is, however, Candida. Normally, a healthy digestive tract and immune system controls Candida and other yeast levels. Only when the yeast is left unchecked and becomes dominant in any of the various parts of the body does it begin to present a health problem. Then it can become a strong, invasive parasite that can attach itself to the intestinal wall, becoming a permanent resident of the internal organs. Once it becomes systemic or chronic, its contribution to poor health and disease can be overwhelming. It commonly produces localized symptoms by invading the mouth, ears, digestive tract, skin, finger and toenails, and the yeast toxins can damage or cripple the immune system, leaving the body predisposed to secondary bacterial and viral invasions, as well.

Left unchecked, Candidiasis is truly a harmful yeast infestation that begins in the digestive system and can slowly spread to other areas of the body. Candida is a hardy, aggressive fungus that mutates rapidly, and it can assume long periods of dormancy, becoming cannibalistic when necessary. A body weakened by a yeast overload cannot remove antigens, and where the immune system is impaired, high levels of yeast toxins can interfere with normal tolerances, which tend to result in sensitivities to specific substances. This creates allergic reactions of varying intensities within the body's own systems. These can include reactions to certain foods, pollen, odors (VOC), chemicals, and various cosmetic products.

Scientists have isolated at least 79 known toxins from yeast, and it is believed that there could be more than 100. After years of overgrowth, the buildup of acetaldehyde toxins can overwhelm the body tissues, and these acid residues are transformed into ethanol by the liver. Acetaldehyde is created as the Candida feeds on the carbohydrates and insoluble fiber in the digestive system. Cell energy is then greatly diminished as magnesium and potassium—two minerals absolutely essential for tissue strength and integrity—are depleted by the conversion of the acetaldehyde to alcohol. Not only is energy reduced, but calcium absorption becomes distorted (due to the imbalance of nutrients caused by the acetaldehyde-alcohol conversion), which can lead to bone spurs, tendon and muscle problems, osteoporosis and immune deficiencies. The less oxygen there is in the body, the more alcohol is produced by this conversion, which can result in symptoms that mimic intoxication and cause disorientation, dizziness, and/or mental confusion, and cause symptoms of panic, anxiety, depression and irritability, as well as headaches. Acetaldehyde can cause excessive fatigue, discourage ambition, and reduce strength, stamina and clarity of thought. It destroys enzymes needed for cell energy and causes the release of free radicals.

When Candida enters the bloodstream it can quickly invade the tissues, causing joint pain, chest pains, sinus inflammation, glandular stress, ear infections, menstrual problems and a host of other complications so common today. The most damaging, however, is the compromise of the immune system. Candida stresses and weakens the immune system, 24/7. As the Candida yeasts and their numerous toxins are being neutralized by the body's immune defenses, the white blood cell count rises and the system is stressed so as to severely exacerbate any pre-existing infectious or immunological conditions. If this type of body pollution is continually generated, eventually the immune system becomes ineffective, setting the stage for severe health problems such as lupus, leukemia, chronic fatigue and other infections throughout the body.

The majority of people who have a Candida overgrowth rarely realize they have it until they become seriously ill. The symptoms are usually numerous, and seemingly unrelated, so the true culprit often escapes both health practitioner and patient. Many of the symptoms are not directly caused by Candida, but will run either parallel to it, or in connection with it. Since Candida weakens the process of absorption, and damages the immune system, the nervous system can be impaired and the production of hormones interfered with. Organ dysfunction and continually compromised health often result, and although an estimated 85% of Canadians and Americans are affected, there is, in my opinion, little suspicion that a Candida overgrowth is at the root of their failing health.

Chapter 6

Candida Symptoms

Although exact combinations and severity of case depend on the individual, a Candida overgrowth can present a wide variety of symptoms. Because of the vast diversity of the following symptoms, they may appear to be unrelated, but in fact all can be caused, either directly or indirectly, by the overgrowth of Candida.

Allergies and Environmental Sensitivities

Many environmentally sensitive individuals blame environmental pollutants and contaminants such as pollen, mold, dust, grass and various foods, etc., for their problems, but these environmental pollutants or elements do not cause allergies; they are only the triggering mechanisms. Candida weakens the digestive system, which affects the total breakdown of amino acids. If any amino acids or even any protein compounds, are not completely digested and broken down, they can be absorbed into the blood and irritate tissues, because the body then sees these proteins as foreign or toxic invaders, and a powerful immune response is initiated. A chain reaction of chemical events occurs, causing an allergic reaction that is triggered by certain foods or airborne antigens. The combination of poor protein digestion and the presence of Candida toxins cause the immune system to become confused, resulting in the chemical or food sensitivities we call "allergies." If there were no

Candida toxins present in the body, or the body had an optimal ability to create enzymes, protein digestion would be efficient and allergies that are bio-chemically created would be non-existent. Nature does not intend that we have allergies and suffer from illness; the Candida yeast overgrowth, which causes these biochemical reactions, is both tangible and correctable.

Another common complaint is environmental hypersensitivity to things such as smoke, auto exhaust, natural gas, and fumes within the home from the carpet, fabrics, and walls. These can cause extreme adverse reactions within the body. This happens particularly when Candida filaments infiltrate the lungs and the sinus membranes, causing serious tissue congestion and inflammation. Candida can be anywhere in the systems of the body, including the brain. Candida weakens the entire body, thereby lowering resistance to all kinds of other diseases and health problems.

Not all individuals suffering some combination of the above symptoms will have a Candida problem, but the likelihood of it is very high. Penetration of Candida filaments from within the intestinal cavity through the intestinal wall destroys the integrity of the membrane system. These little tube-like structures of concentrated fungus literally eat through the intestinal wall, exposing the bloodstream to severe Candida toxins, which are then carried to other parts of the body, where they invade the tissues and organs. The compromised intestinal walls (leaky gut syndrome) can now allow poisonous toxins left by the activities of other microorganisms to also enter the bloodstream, causing a variety of symptoms and aggravating many pre-existing conditions.

Childhood Developmental Concerns

Children are very susceptible to Candida and the associated symptoms. Yeasts and other microorganisms, including friendly bacteria, can be transferred during delivery through the birth canal. If the mother has a yeast overgrowth, or if she is lacking in the friendly bacteria that fight the yeast, then the baby will also have similar problems. A surprising amount of childhood infections and conditions like colds, diaper rash, thrush, ear infections, tonsillitis, colic, constipation and diarrhea are caused by Candida and a lack of friendly bacteria. As children become older, conditions like hyperactivity, aggressiveness, irrational behavior, poor self-esteem, learning dysfunctions and short attention spans can all be contributed to by the Candida conditions within their bodies. Body growth depends on good digestion, and immune responses depend on healthy GI tract function. If Candida gets to the point where it occupies both the small and large intestine it creates gastrointestinal malfunctions.

Fatigue

There are only two elements that can cause fatigue: the increased presence of contaminants and the decreased presence of quality nutrients. If there were no toxic pollution factors within the blood and tissues, and all necessary energy components such as blood sugar, B vitamins, sodium, potassium, magnesium, oxygen, hormones and enzymes were readily available to each and every cell, you would simply have very little fatigue. Fatigue is one of the most common symptoms or complaints of Candidiasis, and can be especially noticed after a night's rest, after eating, and in the middle of the afternoon. An acid body is a tired body. The acetaldehyde and other yeast toxins reduce the absorption of protein and minerals, which, in turn, weakens the ability of the body to produce enzymes and hormones, and in this way, depletes energy.

Because of poor digestion, food is used to feed the Candida, instead of the cells. Consequently, continual hunger and low blood sugar manifest, as the yeast overgrowth deposits its wastes and depletes nutrition. The result is fatigue, poor endurance and weakness. In the majority of cases, Candida can be the primary cause of Chronic Fatigue and Immune Deficiency Syndrome. In short, a Candida yeast infection begins in the digestive system and then spreads to other parts of the body. It robs the tissues of nutrients and causes destructive toxins to enter the blood and poison the tissues.

Hormonal Imbalances

Women are particularly prone to suffer hormonal imbalances, such as an abundance of estrogen and a shortage of progesterone, which gives Candida a superior foothold within the body. Natural conditions such as puberty, pregnancy, (especially multiple pregnancies), and menstruation provoke hormonal swings. These are exacerbated with the use of oral contraceptives or any hormone based medication or therapy.

Candida feeds on progesterone and produces a by-product of prednisone. Prednisone can cause acne, hair problems, softening of the breasts, migraines and depression. Further effects link to fibromyalgia, water retention, heart related stress, lymphoma, mental confusion, PMS and early or difficult menopause.

Hormone based pesticides are in prevalent use, and contact with them will disrupt hormonal balance (not to mention the damage they do to intestinal flora).

The adrenals and the thyroid produce 90% of the body's hormones. Thus, when the adrenals and thyroid are affected by Candida overgrowth, hormone production is affected as well. This will then affect sugar and fat metabolism, heart rate, and stress reaction. Additionally, Candida affects these glands' ability to control sodium and potassium levels, which control fluid levels and the body's ability to exchange nutrients and wastes.

Hypoglycemia and Diabetes

Two major problems in our society today are low blood sugar (hypoglycemia) and high blood sugar (diabetes). Both of these conditions can be caused by Candida yeast if it gets close to or around the digestive organs. The pancreas not only produces enzymes for digestion, but also the hormone, insulin, which allows blood sugar to enter the cells for utilization and energy. The symptoms of too little or too much sugar are devastating to many people, and in some cases Candida can interfere with or destroy the function of the pancreas, as well as the adrenal glands, kidneys, and the liver.

Neurological and Emotional Symptoms

When excessive yeast toxins within the digestive system migrate to the bloodstream, they can cause ailments such as irritability and mood swings, headaches, migraines, a fogged-in feeling, an inability to concentrate, poor memory, confusion, dizziness, and even MS-like symptoms such as slurred speech and muscular coordination problems. The acetaldehyde toxin that is produced constantly by the yeast is absorbed into the body and is converted to alcohol by the liver. Some people who have the symptoms of being drunk are actually showing the effects of Candida and the alcohol by-product that is being circulated within their bodies.

Candida and its poisonous toxins interfere with the production of co-enzyme A. This is a very important part of all chemical processes within the cells and tissues of the body. When co-enzyme A decreases, poor health begins, and conditions without apparent cause, like depression, anxiety, and PMS symptoms, often worsen.

Paranoia, mental incompetence, and a variety of other behavioral disturbances (emotional and psychological) can also be the result of the presence of Candida. Another symptom that can occur is a hypochondriac-type reaction caused by neurotic and emotional instability.

Poor Absorption and Elimination

A diet high in sugars and carbohydrates is actively feeding Candida, which over time can transform the large intestine into a lifeless pipe containing layers of encrusted fecal material and debris. This further promotes Candida and other parasites. The absorption surfaces of the entire GI tract can become impaired, which contributes to malnutrition. Most people are suffering nutritional deficiencies and chemical imbalances so that taking vitamins, minerals, protein supplements, food supplements, etc., make them feel better. However, these are just serving as symptom relief masking a severe condition.

A great percentage of the population is digesting and absorbing less than 50% of what they eat. Without proper nourishment, body tissues cannot and will not heal or generate, the aging process is accelerated, and the body's productivity is compromised.

Gastrointestinal disturbances of all sorts result from poor eating combined with Candida overgrowth, causing indigestion, heartburn, gas, bloating, cramps, intestinal pain, nausea, gastritis, constipation, diarrhea, colitis, ulcers and coated tongue.

Lack of proper nutrition or nourishment appears in the body as brown, green, blue, or black circles around the eyes, sunken eyes, baggy eyes, cracked lips, deformed teeth, waxy appearance on the face, Rosacea, facial spider veins, bruising, and fragile skin, anger, irritability, depression, acidy feeling, fatigue, anxiety, bed wetting, bad breath and more.

If healthy, the intestine contains two to three pounds of more than 900 different strains of friendly intestinal flora. They create an acid environment that stimulates peristaltic action and act as a natural antiseptic. They keep the walls smooth and clean preventing pockets, polyps, impaction, colitis, bloating, diverticulitis and constipation. The micro-organisms synthesize our vitamins without which we risk fatigue and emotional instability and blood clotting.

Reproduction, Respiratory and Skin Conditions

Vaginal infections and menstrual difficulties, impotence and infertility, prostatitis, rectal itch, urinary tract infections, and urination urgency and burning, can all be the result of a Candida overgrowth. Acetaldehyde toxins and the suppression of the immune system can even contribute to the growth of cysts and tumors. The respiratory system is also greatly affected. Problems stemming from low immune resistance such as flu, colds, hay fever, congestion, post nasal drip, asthma and bronchitis, frequent clearing of the throat, habitual coughing that will not respond to anything, sore throats and even earaches are often associated with Candida. The skin responds negatively to Candida with such conditions as athlete's foot, jock itch, skin rash, hives, dry browning patches, cirrhosis, ringworm, rough skin on the sides of the arms, and acne.

Chapter 7

Do You Have A Candida Overgrowth?

Because of the overwhelming evidence that even a one-time stint on antibiotics, or one X-ray or CAT scan, can kill off the gut flora that keeps Candida in check, it is pretty much assumed that anyone who has seen a doctor for an ailment, has, at one time or another, been prescribed antibiotics or X-rays, and therefore, does have a Candida overgrowth. It is the cause of numerous health problems and discomforts for an estimated 45 million-plus people, and it is estimated that nearly half of the world's population has, or will have, a moderate to serious Candida condition at some point in their lives. The reader can also take the test supplied in this booklet, which will indicate whether or not there is likely an overload of Candida.

Being one of the missing links in our medical diagnostic system, Candida can rob the body of its nourishment, poison the tissues with its toxins, and contribute directly or indirectly to the following list of possibly serious conditions:

Aches/Degeneration
Acne
Adrenal/Thyroid
Allergies
Antisocial Behavior
Asthma/Bronchitis
Bad Breath/Body Odor
Bladder/Urinary
Bone Loss
Bruise Easily
Burning Eyes
Chemical Sensitivity
Cold/Shaky Sensations
Colds & Flu

Colitis
Constipation
Deficiency
Depression
Diarrhea
Dizziness
Dry Mouth/Eyes
Dry Skin & Itching
Endometriosis
Epstein Barr Virus
Exhaustion
Extreme Mineral Deficiency
Eyesight Problems
Fatigue (Chronic)
Finger/Toenail Fungus
Food Cravings
Frequent Infections
Gas/Bloating
Hair Loss
Hay Fever/Sinus
Headaches/Migraines
Heart Irregularities
Heartburn
Hemorrhoids
High/Low Blood Sugar
Hormonal Imbalance
Hyperactivity/ADHD
Indigestion
Infections
Inflammation
Inflammatory Conditions
Insomnia
Intestinal Pain
Iron Deficiency
Irritable Bowel Syndrome
Joint Pain/Arthritis
Lethargic/Laziness
Low Blood Sugar
Lupus-Type Symptoms
Malabsorption
Menstrual Problems
Mood Swings
Muscle
No Sex Drive
Numbness
Over & Under Weight
Over-all Bad Feeling
PMS Symptoms
Poor Memory
Premature Aging
Prostate Problems
Puffy Eyes
Respiratory Problems
Skin Rash & Hives
Thrush/Gum Receding
Tingling Sensations
Ulcers
Vaginal Yeast Infection

Chapter 8

How Can I Tell If I Have a Candida Overgrowth?

Candida can be very deceiving and elusive. The best and truest indicators are found in the symptoms. Use the symptom chart and do the math! One can have a blood analysis, but it is accepted that even the labs disclaim at least 15% of the tests will come back as false. Another method that one can use in conjunction with the symptoms chart is the "saliva test." It's easy, and it's free! Any of the reactions or symptoms listed below indicates a positive result. If one does not see the symptoms below, however, it's a "maybe/ maybe not." This is when we look back at the symptoms and consider the probabilities. And remember, if one has had antibiotics or X-rays, he or she is most assuredly a prime candidate, and if there has been no previous treatment for Candida, one can assume a yeast overgrowth is present.

Simple Saliva Test

This simple saliva test can help determine if you're actually battling Candida. First thing in the morning, rinse out your mouth. Then place some natural, non-chemically treated water in a clear glass at room temperature. Work up some saliva, and spit into the glass of water. Closely observe the water for 1/2 to 1 hour, paying special attention to the first few minutes. If you have a Candida yeast infection, you will see one of three, or even all three, of the following:

1. Strings (legs) traveling down into the water from the saliva floating on top.
2. Nasty looking saliva at the bottom of the glass.
3. Cloudy specks suspended in the water.

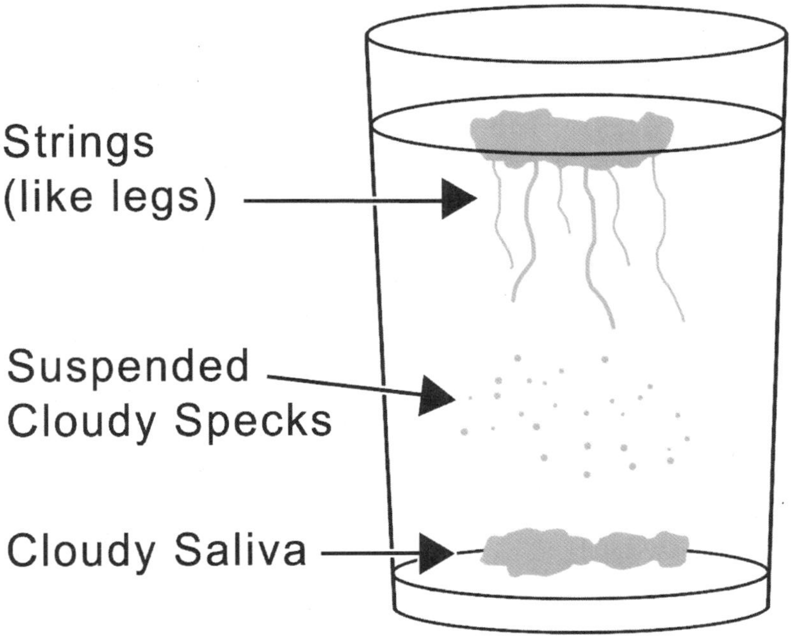

The more you see any or all of these, and the sooner it appears, the worse the overgrowth. If you spit and almost immediately you see stuff falling from the saliva, you know it's advanced. If after 20 minutes you see just a little bit falling from the saliva, you've probably got an overgrowth, but not as advanced. If the saliva just floats on top and the water stays perfectly clear, you probably don't have a Candida overgrowth. Candida concentrates in saliva overnight so it will be more accurate when taking this test first thing in the morning. If you have a fairly extensive overgrowth you will still see evidence of it if taking the test during the day.

Confirm this test by checking to see if you have any of the several Candida overgrowth symptoms.

Overgrowth Symptoms

If you have gotten rid of your symptoms and the test shows clear, you're most likely rid of the Candida overgrowth. (It is very invasive and extremely difficult to get rid of, so if you've had it a long time, or it's quite advanced, it may take quite some time to get it under control, and then some time longer to get rid of it completely.)

Do you...

1. Bloat when you eat?
2. Form gas when you eat?
3. Have acid reflux?
4. Have brain fog?
5. Have sinus or ear infections?
6. Suffer from fatigue for no apparent reason?
7. Have a dry mouth?
8. Have vision that gets blurry, then clear, then blurry again?
9. Have hypoglycemia? (Shaky if meal is missed, sleepy after a meal, sweat during sleep)
10. Have constipation, diarrhea, or both?
11. Have borderline anemia?

12. Have rashes?

13. Have toenail or fingernail fungus?

14. Have vaginal yeast infections or jock itch?

A 'yes' to three or more of these questions indicates fungal overgrowth. Also, take a look at these other typical Candida symptoms to see if you have any of them: Short-term memory loss, persistent drowsiness, headaches, mood swings, dizziness, loss of balance, lack of coordination, ear sensitivity/ringing/itching or fluid in the ears, mucus in stools, postnasal drip, frequent colds, (recurring strep throat, sinusitis or bronchitis), heartburn, nervous irritability, tightness of the chest, white stuff in the throat or coated tongue, bad breath, thyroid problems, depression, sugar disorders. Some of the above are usually worse on damp days. Some disorders that usually are accompanied by Candida overgrowth are: irritable bowel syndrome, autism, fibromyalgia, cancer, diabetes, hypoglycemia, chronic fatigue, Epstein Barr Virus, pneumonia, lupus, acid reflux and hiatal hernia.

Chapter 9

Simple Program

The following suggested dietary and nutritional steps have been very effective in removing Candida overgrowth, viruses, and infectious bacteria. Consequently, when followed, this program will promote and encourage renewed energy, health, and wellness.

The recommended amounts of the following products may vary, depending on the severity of the Candida Yeast overgrowth, the overall condition of the digestive tract, and the extent of the systemic overload of Candida acids, heavy metals, and toxins.

1. Stress reduction!!! Stress reduction!!! Stress reduction!!! Because the delicate balance of the digestive tract is so vulnerable to stress, monitoring emotions and reducing stress from external sources are of paramount importance. Meditation, listening to music, taking walks, limiting the chaos of the outside world: all of these activities are very good at reducing stress. In addition eat red meat, the only natural source of tryptophan and tyrosine, amino acids that are absolutely necessary for stress reduction. Neurotransmitters are released in response to stress, which is our "fight or flight" defense mechanism. Since we no longer fight or flee from the primitive dangers of survival, we now deal with depletion that can only be restored with a sufficient consumption of red meats. Artificial supplementation is not the best or most effective bio-electrical source, although it is probably necessary to supplement, for a time, along with a good source of red meat in order to overcome

acids

the depletion.

2. A biologically and electrically compatible diet is the foundation of our health! *Chapter 10: The Dietary Parameters of Workability* provides the necessary guidelines to follow for optimal health. Additionally, an example of this type of diet is laid out in the book: **Electrical Nutrition**, by Denie and Shelley Hiestand. Today's diet is so devoid of nutrients for true nourishment that it is highly recommended to supplement with an electrically bio-available mineral supplement with an abundance of 60 or more Macro-Minerals, Essential-Trace-Minerals, and Ultra-Trace-Minerals daily. This supplementation promotes sustained energy and supplies the tools to build the healthy digestive system that the other suggested supplements are designed to maintain.

3. Restore the body's "memory" to create enzymes. Poor diet over time results in a cellular "Memory loss" whereby the body no longer knows how to create the enzymes needed for digesting food. The only way we have found to accomplish restoring the "memory" is through the dietary suggestions in this book and supplementation of enzymes. Attention should be paid to the quality, content, and bio-availability of enzyme supplements. While the body utilizes many enzymes for many purposes, it is the digestive enzymes that are required for this purpose. Our best recommendations can be found in the back of this book.

4. Clean, tone, and rebuild the lining of the digestive tract. We find that large doses of Glutamine daily will promote healing and return the natural electrical peristaltic action and function to the GI tract to facilitate proper elimination. Glutamine helps restore the villi which is the home of the microorganisms (friendly bacteria or probiotics), and

where the transfer of our life force actually takes place.

We recommend cleaning and rebuilding the GI tract, the natural chelation of toxins, pesticide residue, and heavy metals, and repopulating the GI tract with soil borne probiotics. Essential to successfully rebuilding the digestive tract, is the addition of New Zealand colostrum, which aids in strengthening the immune system and aids in rebuilding the complete GI tract.

5. Restore healthy balance in the gut flora, by saturating and repopulating the digestive tract with a broad spectrum of both soil borne and colonizing probiotic microorganisms. It has been observed that Bacillus Laterosporus (BOD) affects the overgrowth of Candida Yeast in two ways: it devours the Candida yeast organism, using it as its own food, and it produces inhibitory proteins which limit the growth of not only the Candida yeast, but other pathogenic organisms like E.Coli (Escherichia Coli), Salmonella, Streptococci, and Staphylococcus, as well. A broad spectrum of probiotic microorganisms will support the benefits of BOD) as well as keep all microorganisms at beneficial levels of development.

6. OxyLift®: It's believed that all forms of illness can be traced to two basic causes: too many toxins in the body and too few nutrients reaching the cells. These are carried in and out of the cells in water based fluids. Drinking even large amounts of water does not ensure the body is hydrated sufficiently. Most water has a surface tension too resistant to absorption for it to do more than wash through the body, instead of saturating, cleansing, and bringing nourishment to the cells. OxyLift® addresses these vital health issues at the same time, and with remarkable thoroughness. Its unique structure oxygenates and feeds the cells: cleaning and tun-

ing up the body’s systems throughout the day.

Chapter 10

Dietary Parameters of Workability for Maximum Health and Vitality

The following suggested protocol has been developed to enhance the correct parameters necessary for optimal digestion, reversing physical degeneration by natural chelation, and rebuilding the body on a cellular level, repopulating the digestive tract with microorganisms and providing some of nature's most powerful, electrically and biologically available health promoting elements.

The body runs on electrical energy and has an electrical system. Quantum Physics verifies that there is nothing but energy in our universe. In his book, **The Genie in Your Genes**, Dawson Church, PhD, describes the body's electrical system components as all the muscle tissue, connective tissue, cell membranes, the myelin sheath of the nerve cells, the rod and cone cells in our eyes, and the DNA of our cells.

Contrary to traditional nutritional understanding, we eat to accomplish an electrical transfer of energy from food to our cells through the process of digestion. Vitamins, amino acids, and the like are managed bio-elementally in an intricate, cooperative electrical process of exchange between our digestive tract and the microorganisms within it. According to Denie Hiestand in his book, **Electrical Nutrition**, and Saul Liss, PhD, the body obtains and sustains its energy or life force by ingesting available amino acids and minerals. Nourishment from this perspective requires certain rules about what to eat, what not to eat and what to eat together, or not.

While correct dietary elements and food combining are paramount to your good health, the fact is that regardless of the quality of the food you are eating, your diet will be deficient without supplementation. Replacing the vital components lost in our degrading environment will provide an electrical nutritional balance to restore your body's innate ability to be well and vibrant physically, mentally, emotionally, and spiritually.

Diet

Using the "rock to rot" theory as outlined in detail in the book, **Electrical Nutrition**, by Denie Hiestand, is the best approach when in doubt about what is good for you to eat. This means that food which will rot within seven days if left to set out on your counter will break down in your stomach's digestive juices within the three hours your stomach has to accomplish fermentation. Food that has not been sufficiently "rotted" or fermented in your stomach requires your intestine to attempt to process unprepared material, preventing the natural function of the small intestine.

Your body prepares itself to eliminate upon rising or shortly thereafter during the morning hours before waking. Eating fruit for breakfast is light enough so as not to slow elimination while the fructose is a natural stimulant. We recommend eating fruit in the morning along with your mineral supplement and your pro-biotics. Heavy cream or whole milk yogurt with a bowl of fruit is beneficial as the amino acids in the cream stimulate peristaltic action.

The Rules

The amounts and times of day the rest of your meals are taken are very personal and optional as long as you keep within a few basic guidelines and allow your biology to guide you to knowing its needs. The basic rules are as follows:

Do not eat grains, nuts or seeds. Period.

We simply cannot break open the cellular structure of insoluble fiber through chewing and the 3 to 4 hours of our digestive process in the stomach, making the nutrients in the grains and seeds completely unavailable to us. Since this is so, eating grain requires excessive insulin and gastric juice production in the effort of digestion which creates a very high acid environment, acidosis, causing an optimum opportunity for Candida overgrowth as well as several other equally disturbing and undesirable conditions to manifest and flourish.

Do not eat starch as in potatoes, squash, or legumes with animal protein.

Digestion begins with the sight, smell, and thought of the food we are about to eat so that saliva and digestive fluids are ready when the meal begins. Protein and starch require different solutions for fermentation. Eating them at the same time causes opposing solutions to mix and neutralize, stopping the digestive process and causing buildup of methane and other harmful toxins severely taxing the liver and lower GI tract.

Do not eat fruits and vegetables together.

As with protein and starch, fruits and vegetables require different digestive parameters for fermentation.

Stay away from refined sweeteners.

High fructose corn syrup, Sorbitol, Splenda, Xylotol, Aspartame, white granulated sugars, and the like are simply poison to the system. Stevia, honey, sorghum, and molasses are acceptable in moderation.

Do eat as fresh and raw as possible, including your meats and eggs.

In my clinical experience a simple, basic diet is easy to understand and demands a willingness to change views and habits.

Human beings need to eat protein often in the form of fresh meat and fat (lamb, beef, fish, pork, fowl, eggs, etc.), as raw as possible, for the electrical and biologically available amino acids, large amounts of minerals, and soluble fiber. Current, generally accepted, cautions regarding too much fat in the diet do not apply in a grain free dietary lifestyle.

Some vegetables and especially fruits are a good source of soluble fiber, which feed the microorganisms in our intestinal tract. Yes, it does feed Candida. However, a healthy intestinal tract does not have Candida overgrowth, while Candida does have its place in a healthy gut.

Use whole milk, cream, cheese, and real butter often, preferably raw.

Lactose intolerance, as most other allergies, is a simple result of a gut imbalance, which can be restored as the body returns to its natural state by supplementation and this simple and easy diet protocol.

Stay hydrated.

Natural, clean water is nearly impossible to obtain. Treated water is toxic to the body. Filtered water ends up with a strong surface tension making it difficult for the body to absorb, requiring excessive amounts of intake that leach precious minerals and strip delicate tissues. In the Body Electrician Clinic we use filtered or fresh untreated well or spring water with a little of our moist sea salt, probiotic solution and OxyLift®, to provide excellent hydration, mineralization and lessen the thirst.

What to Supplement

With a healthy population of micro organisms in the digestive tract, sufficient mineral intake and a balanced, electrically bio-available diet, conventional vitamin and dietary supplements are rarely, if ever necessary. The supplements we recommend are the elements that are missing from the food chain as a result of the degradation taking place in our environment and thus in the food chain.

Probiotics

Probiotics are microorganisms that protect their host and prevent disease.

Conventional and even organic agriculture, shipping and packaging methods destroy most microorganisms needed for health. Probiotics in the gut keep it clean through natural chelation of toxins and heavy metals and provide 80% of natural immunity and 100% of nourishment recognition for the body.

Probiotics synthesize all needed vitamins, performing electrical recognition of amino acids and minerals. Soil born or transient, non-colonizing probiotics need to be supplemented continuously, because they are intended by nature to occur in almost every bite of food, but are not present in today's food chain, organic or conventional.

Minerals

Due to the degradation of our food supply by over farming and pollution, supplementing with a wide array of bioavailable, colloidal phytogenic minerals, carrying a natural negative electrical charge is paramount to health. Minerals are necessary for all electrolysis in the body, which is important for cellular regeneration and chelation.

Killing Pathogens

Transitioning from dietary choices and life style that are

less than optimal for vibrant health requires killing off and re-balancing the pathogenic over load that invariably develops throughout the entire body. Conventional and allopathic methods, and even in the natural health arena, accomplishing this is a near impossibility. Using a powerful pathogen killer like sodium chlorite could be a good first step. However, the help of an experienced practitioner with a record of success helps immensely. Pathogens are cunning survivalists and can mutate or go dormant to avoid effective elimination. Applying a consistent approach and understanding how and when to alter the remedies produces the best results.

Again, this suggested protocol is offered as a result of my personal and clinical experience. These are general suggestions for anyone interested in enhancing their own health and wellness. This basic methodology allows for significant personal choice as we consider it appropriate for people to manage themselves personally, mentally, physically, emotionally, and spiritually.

Cleansing Reactions

In the process of self improvement through health enhancement, changes will necessarily be made in diet and supplementation which have an effect on bodily functions, systems, elimination, etc. There are reactions that take place as the body rids itself of toxins, especially in the initial phase of health enhancement. These reactions are often referred to as cleansing reactions, but are the same as or similar to what are also referred to as healing trauma, healing reactions, healing crisis, detoxification and possibly others.

According to Jameth Sheridan, N.D., symptoms of a cleansing reaction can include lethargy, temporary all-over muscle aching, mucous or other discharge, a coated, pasty tongue, light-headedness upon standing, headaches and flu-like symptoms in general. A cleansing reaction/healing crisis may take the form of old symptoms that have been previously suppressed. You may think you are sick. However, this cleansing process is absolutely essential to achieve health and should not be suppressed with drugs. The mainstream medical establishment has completely closed its collective mind to cleansing and assumes that any symptoms displayed by the body are caused by a microbe, so antibiotics are usually prescribed even if they think the causative agent is a virus. Medical antibiotics have no effect on viruses, other than weakening the immune system so viruses can replicate more. Antibiotics and other drugs suppress cleansing and the immune system as well.

In a conventional system of health and diet, colds, flu, hives, fevers, diarrhea, skin irritations and the like are spontaneous occurrences as the body purges itself of the build-up of mucous, pollutants, impurities, parasites, etc. Yes, there may be a virus or bacteria involved which usually gets the blame, however it is the build-up that allowed the virus or bacteria to enter. Additionally, the virus or bacteria is then a catalyst for purging an unhealthy build-up in the body. This is a cyclic pattern in conventional diet and healthcare, which does not occur, in a healthy, balanced body.

Similar reactions can be induced in this first phase of health enhancement. Feeding the body what is needed, and has been missing, now provides the ability to more readily purge impurities. This is especially true if there has been conventional medical treatment involved. Medications typically leave a toxic residue for the body to clean up, usually stored in fat until the body is able to deal with it. Advanced diagnostic procedures such as X-rays, CAT scans, anesthesia, and nuclear testing destroy much of the intestinal microorganism population, which in itself can cause problems.

So, these cleansing reactions are really healing events and although they don't feel very good they are not negative events. Looking at it this way there is the potential for feeling worse before feeling better, which is necessary and beneficial. The good news is that as healing continues and the body gets stronger and cleaner these cleansing reactions diminish and after this initial phase common complaints and chronic conditions are usually significantly lessened.

Symptoms can be reduced through a less aggressive approach, taking more time with dietary changes, and the addition of the correct supplements. Increasing water intake flushes and cleans the system more swiftly. Using herbs and natural remedies for symptom relief are helpful in this stage.

Most importantly, relax and stay calm. Making these positive changes is good for you, so know you are giving yourself goodness. Nurture yourself with rest and confidence. Given sufficient time you will feel better than perhaps ever before.

Chapter 11

An Alternate Perspective

The good news of the unconscious conspiracy is that although it bears a lot of incriminating evidence, it is not so much a conspiracy as it is a matter of consciousness. When we can open our eyes sufficiently to really see what is at the core of impact on life on our planet, we can grow in the awareness that our real and lasting answers have been in plain sight all along. They have been unconsciously overlooked through our psychological pull toward the path of least resistance, convenience, and the thrill of modern technology. Wanting to continue to increase and proceed has kept us from taking the precious time and space to slow down and really assess the price we are paying, because it is a price that does not show up in the accounting books until a moment like now, when environments—both those within our bodies and those in which we live—have provided the discomfort we have needed to wake us from this sleep.

Now that we are waking, connecting the dots between the issues at hand brings the clarity of a larger picture and a simpler problem, or focus on the simplicity beneath the problem of restoration.

It will take a collective commitment to interrupt and transform the world of detrimental practices. They are massive and far-reaching and are fortified with their own longstanding habitual resistance and self defenses.

However, the alterations we can make using knowledge

of real solutions based on nature's own recycling forces are powerful indeed. There is no force on Earth more powerful than Nature herself, and when we align with her curative remedies we hold the hand of the inevitable: Life.

Earth not only has the capacity, it is inevitable that it will continue to evolve with or without us on her skin. It is, in fact, unrealistic and arrogant for us to consider otherwise. The real question is whether we are willing to adjust to nature and remain here or to continue unconsciously creating an environment we cannot survive.

Bioremediation

Microorganisms are nature's recycling agents. They have already been proven to be the most effective agents used to clean up the worst possible disasters in history.

A quote from "Bioremediation Methods for Oil Spills", Acres USA magazine 2006:

"The biodegradation of petroleum in the marine environment is carried out largely by diverse bacterial populations, including various Pseudomonas species. The hydrocarbon-biodegrading populations are widely distributed in the world's oceans; surveys of marine bacteria indicate that hydrocarbon-degrading microorganisms are ubiquitously distributed in the marine environment. Generally, in pristine environments, the hydrocarbon-degrading bacteria comprise $< 1\%$ of the total bacterial population. These bacteria presumably utilize hydrocarbons that are naturally produced by plants, algae, and other living organisms. They also utilize other substrates, such as carbohydrates and proteins. When an environment is contaminated with petroleum, the proportion of hydrocarbon-degrading microorganisms increases rapidly."

Carl H. Oppenheimer, PhD

Bioremediation refers to any method that uses microbes (microorganisms) to recycle organic materials and sequester inorganic ions. Because the primary responsibility of microbes is to recycle organic material, they must be present in sufficient quantities and diversity in order to do this. Under careful controlled conditions, bioremediation can be a practical and cost effective method to remove hydrocarbons from contaminated surfaces and sub-surfaces. Bioremediation is the basis for composting of human and industrial waste recycling.

Oppenheimer's early career led up to determining ways to clean up hydrocarbon pollution using microbes. He is known to be one of the world's founding fathers of bioremediation. He was instrumental in persuading Texas to allow for the first bioremediation open water testing in the U.S. in which he used his manufactured microbial consortium known as the Oppenheimer Formula. This product is made up of safe aerobic and microaerophilic microbes, specific for breaking down hydrocarbons. Through the use of his bioremediation products and techniques, he established ways to clean up polluted waters and restore these diverse ecosystems. He founded Oppenheimer Biotechnology, Inc. in 1990, located in Texas, which manufactures bioremediation products that have been used throughout the United States and the world. His products are able to successfully and effectively clean up a wide range of contaminants in the environment.

When a big oil spill makes headlines, however, you probably will not read that Carl Oppenheimer or someone he's educated has been hired to clean it up. Less effective dispersal technology is still the industry standard. "We can't do it on the ocean . . . (T)hey have a physical process they are heavily invested in – booms and skimmers and so on – and we'd put them out of business." (Source: "Bioremediation: Using Microbes to Clean Toxic Environments", ACRES USA, magazine 2006 "A Voice for eco-agriculture") An excerpt from the article.

Oppenheimer pours a few tablespoons of motor oil into a porcelain bowl, then spoons some grayish powder onto it and stirs the mess around a little bit. Fizzing occurs. . . "It turns into fish food. . . The fish jump for it and gobble away." The reporter is google-eyed. "What is it now?" he wants to know. "Fatty acids," Oppenheimer answers, "from petroleum product to fish food in five minutes."

Think about it: since Oppenheimer's microbial arrays can eliminate most pesticide residues, the time it takes to effect a transition from conventional to clean farming could be reduced dramatically, far below the regulation three years.

The Environmental Inquiry

In a non-polluted environment, bacteria, fungi, protists, and other microorganisms are constantly at work breaking down organic matter. What would occur if an organic pollutant such as oil contaminated this environment? Some of the microorganisms would die, while others capable of eating the organic pollution would survive. Bioremediation works by providing these pollution-eating organisms with fertilizer, oxygen, and other conditions that encourage their rapid growth. These organisms would then be able to break down the organic pollutant at a correspondingly faster rate. In fact, bioremediation is often used to help clean up oil spills.

Microorganisms can and will break down, dissolve, recycle, and renew material wherever there exists an opportunity for them to do so. Some of the material that these microorganisms can break down and render bio-compatible and non-toxic are crude oil, petroleum based compounds such as petro-chemical products, motor oils, gasoline, heating oil, and diesel fuels, solvents, wood preservatives, waste products, most pesticides, DDT, PCB's, PCE, TCE, greases, vegetable oils, pulp by-products, brake and hydraulic fluids, heavy metals, salts, highly insoluble compounds, and some synthetic oils.

If the environment they grow in is already in a balanced stasis, they will not survive or they will remain within parameters appropriate to the existing stasis. However, if there is a lack of stasis, they will proliferate, often one microorganism feeding off of another, until a new stasis is achieved. Life renews life in this way.

Ultimate Benefits

Microorganisms can break down rocks in the soils, releasing minerals for plant root uptake and enhancing the nutritional values in foods. The rich supply of minerals and microorganisms shortens germination time and produce healthier, stronger plant life. In this way the food chain begins to rebuild.

Animals and people feeding on the increased nutrition and microorganisms in plants and soils now receive nature's resources for their health, immunity, and the ability to thrive in a less than optimal environment.

Meats and dairy products from animals raised on mineral rich pastures provide enhanced sources of amino acids, minerals, and fatty acids. Consumed along with microorganisms, we now have humans reclaiming their internal environments and vibrancy.

Enhancing healthier internal environments enhances overall human health, mental clarity, energy, immunity, and the ability to thrive in a less than optimal external environment. Generational nutritional decline reverses, and childhood disease, maladies, birth defects, etc. begin to disappear.

Meanwhile, the external environment steadily improves with the addition of microorganisms everywhere, as we collectively tackle reversing the gigantic avalanche upon us.

Living Powerfully

Living life powerfully begins with the individual. From there it spreads to the environment of the individual, the community, the town, the garden, the city, the farm, the state, the country, the planet. With clarity and conviction, understanding what really works and being unstoppable in one's commitment to personal health is the ultimate simple answer to the unconscious conspiracy. Personal health begins within one's body, which really begins with the soil and follows a natural progression from the soil through all levels of life with the natural outcome an interest in and a commitment to living on a clean and healthy planet. From the communities of microscopic life in the soil to human communities, be they domestic, farming, business, corporate, international, or governmental, all must coexist, co-operate, clean up, and thrive.

Living in community when the community is harmonious and supportive of true workability opens infinite possibility for transformation on a planetary level. Co-operative systems and living provide huge benefits in this regard.

Almost every locality already has in place food co-operatives and most have farmers' markets. A short trip out of town can almost invariably find a dairy farm producing raw milk, cheese, butter, yogurt, and free range if not organic eggs. That same trip could very likely identify a butcher who takes humane care of his animals, making sure they are free of toxic elements that affect their meat and pastures them on healthy, natural grasses. Many areas, even in the cities have available plots for personal gardening. Growing your own food is a great way to be sure you are getting foods grown with minerals and microorganisms to provide real nutritive factors. All of these things are significantly more rewarding when practiced in a community dynamic with others of like mind. Even a flowerpot garden on a second floor deck, if fed appropriately will provide more nutrition than much of what is found in the grocery stores.

We believe that the human being is innately a gregarious, caring, family oriented being. The isolation so prevalent among our societies today is not natural or nurturing for us. Returning to an atmosphere of the ancient values of the family, the clan, the small community, does not require relinquishing the individuality. In fact, conversely, the security of acceptance and neighborly love do much for happiness and progress. Sharing ideas, responsibilities, and even equipment supports the individual in many ways.

Confident, healthy communities nurturing one another, living and growing in the concepts of health and mutual acceptance and support, assure a solid future for generations to come. Children growing in such an atmosphere learn to believe in positive values of health, wholeness, safety, and unity, which provide the largest health benefit possible. Living in love, nurture, and acceptance, free of fear, anger, and stress is proven to be the most impactful element of good health available.

Appendix I

The Amazing Benefits of Bacillus Laterosporus (BOD)

Testing of the product at the University of California, Davis revealed that Bacillus Laterosporus BOD inhibited the growth of several common pathogenic bacteria. Further testing revealed that Bacillus Laterosporus BOD is extremely effective in controlling a wide range of disease causing organisms. For example, the organism killed 99.8% of a culture of Salmonella in seven days. Salmonella is a bacterium that is often found in raw poultry products. It is recommended to fully cook all poultry products including eggs to avoid the possibility of Salmonella poisoning. Once Salmonella is established in the intestine it can be extremely difficult to control and can be life threatening.

In the past few years the bacterium E. Coli has made the news because it had been found in hamburgers at some well-known fast food establishments where the meat was possibly not cooked long enough to kill the bacteria. Bacillus Laterosporus BOD killed 97% of E. Coli in 21 days. This is significant because E. Coli reproduces extremely rapidly and is relatively resistant to modern antibiotics.

The most significant result was obtained with the yeast/fungus Candida; a 96.4% reduction was achieved in just 28 days. This remarkable result is important because Candida, a form of yeast, is so difficult to eradicate once it has taken over the intestinal tract. Long-term Candida infection can lead to other health problems, and has been implicated in many of today's newest and least understood diseases such as Chronic Fatigue Syndrome, and Epstein-Barr virus infection.

Another important benefit of Bacillus Laterosporus BOD is that, unlike many other Candida treatments, it does not harm natural lactobacilli in the intestinal tract, but rather helps their survival. These friendly bacteria are necessary to maintain a healthy environment in the intestine, which leads to better digestion and absorption of nutrients from ingested food.

Food in the intestine that has putrefied because of incomplete digestion forms tacky clay-like clumps that stick to the wall of the intestine, and may interfere with nutrient absorption. Bacillus Laterosporus BOD assists in breaking down the accumulation of the intestinal clay-like clumps, restoring the intestinal lining's nutrient absorption ability.

The fact that Bacillus Laterosporus BOD kills certain pathogenic bacteria has been proven. Evidence also exists that it helps bolster the immune system. Studies indicate that, in response to the presence of Bacillus Laterosporus BOD, the body produces and stores antibodies that can be used when needed to fight off pathogenic organisms. Preliminary research also indicates that this bacillus may also increase natural interferon, an immune booster that helps fight cancer and infections. Bacillus Laterosporus BOD taken daily is an excellent preventative measure against infections.

It is interesting that Bacillus Laterosporus BOD may have provided protection against disease centuries ago, when it was plentiful on the earth. As we have seen, theories have been postulated on the reason for its decline. Since the beginning of the industrial revolution man has added huge amounts of chemicals into the environment, which may have contributed to the diminution, and almost total extinction of, Bacillus Laterosporus BOD.

Appendix II

Proprietary Products

Advanced Oxygen Therapy

- Our most powerful pathogen killer

70+Minerals

- A perfect balance of electrically Bio-available minerals

100+Probiotic Solution

- Over 100 soil borne strains of probiotics

Transcendental GI Cleanse

- 230+ more soil borne probiotics, plus proprietary ingredients

Prozymes

- Digestive enzymes blended with probiotics

DigestAssist

- A fermentation aid for digestion

Sal de la Luz

- The cleanest mineral rich moist sea salt available

EMLC

- A natural source of bio-available amino acids

Colostrum

- Nature's first food for enhancing and restoring health

Note: Proprietary products are made available because they are not found elsewhere in the marketplace. Our protocols use a number of products that can be found in the marketplace as well.

For more information, visit the websites:

www.bodyelectrician.com

www.environotics.com

www.vibrantbees.com

www.vibrant-earth.com

www.fodderproduction.com

www.soil2soul.com

Or call 610-417-7248

Appendix III

The Body Electrician Clinic

Who We Are

With Dr. Wil at the helm, we at The Body Electrician Clinic are a core group of folks who live in a commitment that life on Earth will one day, once again be natural, pristine, and as conducive to vibrant life as it once was. Wil and Delanne are our mentors, in a continual search and a never ending conversation for wellness. We are here and will remain as long as there are products to provide and people in need of them. That is to say, until the environment once again holds and renews the elements we must put in bottles and jars because they cannot exist outside, which we need inside our bodies in order to be healthy.

What We Do

The Body Electrician Clinic provides a number of approaches and modalities for supporting and restoring nature in the human being. Dr. Wil is relentless in his efforts to find and provide workable solutions, products, and modalities to discover where and why nature has become out of balance and the ways to reverse that trend. He honors the individual who lives in the body and teaches the restorative value of not only healing in the body, but the power of the mind and its impact on health.

We consult, teach, and learn. We develop proprietary blends to sustain the physical being. We talk and we listen, supporting the concept of community. We hold the position, like a stake in the Earth, that nature knows best and our best hope is to align with nature.

How We Do It

At a distance, we have facilitated fantastic results helping people to reverse and heal from all manner of diseases. There are several effective ways we accomplish this:

Health Assessment - After filling out an extensive health assessment form, a client spends an hour or longer in a consultation with Dr Wil via Skype or telephone. The client gleans an understanding of the state of their health from Dr. Wil's unique perspective. This includes personal suggestions and recommendations for enhancing wellness.

GSR Hand Scan - Taking advantage of the electronic age, we have developed a program for bringing your health care to you using state-of-the-art technology to read up to 18,000 frequencies throughout the body in the comfort of home.

90 Day Vibrant Health Program - A tried and true combination of supplementation and dietary structures, along with consultations designed to transition the client's body and health from a toxic, conventional state, to one of natural stasis and function.

At the office we are able to work more personally. For example:

Vibrational Re-Set - A process that realigns the electrical circuitry that runs through the body meridians, reconnecting interrupted and broken circuits, releasing energy blocks that can block healing and hold traumatic emotional scars in place. This is extremely impactful and beneficial.

Psyche-K - A cutting edge kinesiology process for removing the conflict, or disagreement, between conscious, chosen, beliefs and hidden embedded, subconscious beliefs that are often not even known to the individual. Often a life altering experience.

Medicinal Mud - Earth energy in the mud packs serves to release the biofilm that builds in areas of injury or surgery. It also detoxifies and allows the body to heal.

You can reach us at:

The Body Electrician Foundation
2430 Butler St. Suite 105
Easton, PA 18042

610-417-7248

wil@bodyelectrician.com

www.bodyelectrician.com

About the Author

Wil Spencer, D.PSc, VMSP

Researcher

Naturopath

Member, Pastoral Medical Association

wil@bodyelectrician.com

Wil Spencer, D.PSc, VMSP, Author, Naturopath, and Researcher is an advocate for Nature. His work extends into multiple areas of investigative research and education regarding human and environmental health relative to identifying and restoring the degraded conditions of natural homeostasis, most often due to industrialization and the misunderstanding of the importance of microorganisms. Operating from the conviction that nature has evolved with an innate wisdom of energetic signatures that influence and interrelate with one another to form a natural, healthy terrain of existence and co-existence, Wil believes all conditions of illness and imbalance are correctible through restoring the elements of nature's design. Wil has applied this conviction successfully in his work reversing and healing conditions of human health, animal health, plant health, including reversing the affects of genetic modification, and increasing the nutritive values of food, reversing the many issues of honey bee health, restoring vitality to soils and compost, water purification, and air quality.

Wil is the resident clinic Naturopath at www.BodyElectrician.com, where he educates and supports his clients, worldwide, to reverse and heal all manner of disease and restore themselves to Vibrant Health. Wil's approach to sustaining vibrant health is based on a paradigm of dietary lifestyle along with nature based supplements. He provides the unique proprietary supplements he recommends through his online product store at www.Environotics.com, where he also makes available his supplements for plants and soils, animals, honey bees, and remediation.

Current endeavors include a committed outreach to the Gulf of Mexico coastal residents for whom Wil has contributed countless hours, consulting with victims of severe chemical poisoning, and products, for some who are financially devastated. His program for recovery has shown verifiable benefit to the health challenges presented by open bacterial wounds, abnormal bleeding, respiratory conditions, fatigue, neuropathy, mental confusion, and depression. Wil has an abiding personal mission to make a difference for the innocent victims of the Gulf Coast disaster to regain their compromised health.

Likewise, Wil has a passionate concern for the victims of Morgellons and Lyme. While his work with clients has only just begun, he is researching the possible connection between these conditions and geo-engineering practices as well as GMO's (genetically modified organisms), and, his therapies have already achieved some intriguing and promising results.

Wil's associate membership in the Morgellon's Research Group was aligned with his interests in air quality and geo-engineering as well as human and environmental protection and planetary health.

Wil speaks at seminars, on webinars, YouTube, web radio interviews, and holds classes throughout North America on topics of Genetically Modified Food and Organisms, Pesticide use, Soil and Environmental Health, Diet and Food, Sustainable Farming Practices, Green Building Techniques, How to Be Healthy in This Chemical Soup of an Environment, Spirituality, Consciousness, etc. He has a long history of humanitarian involvement, having served as a member, lobbyist, and officer for several environmental groups, namely Minnesota Pesticide Awareness (MnPA), Pesticide Action Network North America (PANNA), Minnesota Natural Health Coalition, Healthy Legacy, Organic Consumers Association, Food Alliance, Institute for Responsible Technology, and Natural Solutions Foundation. Wil has served in air and water quality testing and monitoring and has been certified as an Air Quality Sampler. Additionally he is a long standing member in the Weston A Price Foundation.

Wil has accomplished several studies successfully working with commercial honey bees, restoring them from near collapse to full recovery, accomplishing pollination and producing a honey crop in the same season treatment began.

Wil's background and training began with an intimate understanding of agricultural practices and the industries associated with food processing and distribution. Having severe personal health challenges as a result of being raised in a conventional agricultural environment. Wil's quest for self cure resulted in an extensive, unconventional educational process producing in him a deep understanding and appreciation for nature and the ability to assist others in their own quest for wellness.

Discovering the medicinal qualities in plants, Wil became a Master Herbalist. He went on to learn Endorphin Therapy and then a progression of alternate healing modalities. Among them are; Reiki Master, Photonic Therapist, Touch for Health, Quantum Touch, 5 specialized Massage Therapies, Emotional Freedom Technique, Trigger Point Therapy, Acupressure, Reflexology, and Shiatsu. Wil has earned diplomas from the International Institute of Vibrational Medicine in; Energy Awareness, Emotional/Mental Mastership, Energy Kinesiology, and Vibrational Medical Science

As a member of the International Assembly of Spiritual Healers and Earth Stewards, Wil is; a Diplomat of Earth Stewardship, a Certified Spiritual Healer, and Ordained Minister. He has received Native American shamanic apprenticeship. Wil is a contributing writer for the Genesis 2 Church of Health and Healing.

Wil is also a gifted and innovative healer, using his hands to release and correct the flow of electrical energy along the body meridians, where embedded traumatic interruption can hold pain, illness, disease, and emotional upset within the being, providing a modality he calls a Vibrational Re-Set.

Wil lives and breathes his quest for truth in the endless inquiry, "why?" Distinguishing between symptom relief and real sustainable, natural cure is his primary avocation.

Acknowledgements

I would like to acknowledge the thousands of clients who have shared this path with me.

A special thanks to Delanne Walts, Georgie Offrell, Billy Pfeffer, Josh Albanese, Pat Kedrowski, Chief Strong Bull, my parents Darol and Carol Spencer, and all the other individuals who have been a part of this journey of mine.

My sincerest gratitude to Denie and Shelley Hiestand, these two beautiful people have extended to me a love and sharing that is rare in this world today. I first met Denie after listening to him speak in a seminar in Georgia about agriculture in New Zealand, food, and the body's electrical system. I knew then I was blessed. Denie and Shelley's classes have allowed me to open my heart. They have helped me see not only the beautiful person I am but that everyone has the potential of being beautiful from the heart outward.

References

I would also like to acknowledge the following authors and researchers, whose relentless dedication provided the material referred to in this writing:

Burkholder, P.R. & McVeigh, I. (1942, May 23). Botany: Synthesis of Vitamins by Intestinal Bacteria. Vol. 28 (p.285.) Retrieved on July 27, 2016 from: http://www.pnas.org/content/28/7/285.full.pdf

Carlsen, R.H.G. (2001). The candida yeast answer: The ultimate all natural anti-fungal program for complete and permanent yeast removal. Provo, Utah: Candida Wellness Center.

Church, D. (2009). The genie in your genes: Epigenetic medicine and the new biology of intention. (2nd Edition). Santa Rosa, Ca.: Energy Psychology Press.

Corsello, S. & Gallup, J. (n.d.) Bacillus Lateroporus bod: A revolutionary new approach to control candida and yeast infections, Nature's secret is now revealed for balancing the intestines, enhancing the immune system, and maintaining good health and longevity. New Zealand: Living Foods Lifestyle.

Crook, W.G. (2005). The yeast connection: A medical breakthrough. New York, N.Y.: Vintage Books of Random House, [(Recheck -1989) Handbook 2007]

Heinrich, E. (2004). The root of all disease. Tulsa, Oklahoma: The Rockland Corporation, TRC.

Hiestand, D. & Hiestand, S. (2001) Electrical nutrition: A revolutionary approach to eating that awakens the body's electrical energy. New York: Avery, of Penguin, Putnam Inc.

Keith, L. (2009). The vegetarian myth: Food, justice and sustainability. Flashpoint Press: Crescent City, Ca.

Liss, S. & Liss, B. (1999, September). Modulated electrical energy stimulators. Presented at the American Academy of Pain Management Conference. Las Vegas. Paterson, New Jersey: MEDI Consultants. Retrieved from: http://www.ensrmedical.com/wp-content/uploads/ 2014/01/03-Liss-Gathering-Effect.pdf

Lorenzani, S.S. (1986). Candida: A twentieth century disease. New Canaan, CT.: Keats Publishing.

Martin, J. M., & Rona, Z. P., (2000). Complete Candida Yeast Guidebook (2nd Revised ed.). New York, NY: Three Rivers Press.

Monastyrsky, K. (2008). Fiber menace: The truth about the leading role of fiber in diet failure, constipation, hemorrhoids, irritable bowel syndrome, ulcerative colitis, crohn's disease, and colon cancer. U.S.A.: Ageless Press.

Murray, M.T. (1997). Chronic candidiasis: Your natural guide to healing with diet, vitamins, minerals, herbs, exercise, and other natural methods. New York, N.Y.: Three Rivers Press.

Oppenheimer, C.H. (2006). Bioremediation using microbes to clean toxic environments. Tulsa, Oklahoma: Acres U.S.A. Magazine.

Tompkins, P. & Bird, C. (2002). The secret life of plants: A fascinating account of the physical, the emotional, and spiritual relations between plants and man. New York, NY.: Harper Collins Publishers.

Trowbridge, J. P, & Walker, M. (1986) The yeast syndrome: How to help your doctor identify & treat the real cause of your yeast-related illness. New York, NY: Bantam Books.

End Notes

I would also like to acknowledge the following authors and researchers, whose relentless dedication, and conversations with leaders, provided additional material referred to in this book.

McGowen, G. (2010-2014). All conversations, research, and papers generously shared.

Steffy, D. M. (n.d.) All research and papers generously shared.

O' Donnell, B. & O'Donnell, W. (n.d.) All research and papers generously shared.

Testimonials

I have suffered over 20 years under near constant doctor's care including countless rounds of antibiotics and finally steroid injections each month for the last 4+ years. I have had a total of 11 sinus surgeries, one of which removed my frontal (above my eyes) sinus cavities completely. The chronic infections, drainage and severe discomfort continued and my sinuses were plagued with polyps and the steroid shots were very damaging to my health.

When I heard Wil Spencer speak I became interested in finding out if his new approach to health and wellness could make a difference for me. I followed his basic diet, and added his supplements to my daily routine.

I began to feel better rapidly and within 2 months my doctor pronounced my polyps gone. The sinusitis had also abated so that I did not need my monthly steroid injections. I was elated.

Then in another month or so I began to experience a return of the sinus infection symptoms. I called to check in with Wil before going to see the doctor. Wil suggested the antibiotics would undo much of the improvements I have already made and asked me to try another approach first.

Which did indeed resolve the infection and "catching a cold" feeling. Recent blood tests have revealed my body is totally clear of any staph infection or MRSA.

Roger P. Sheridan, MT

In 2000, I was in a car accident that created soft tissue damage in my back. I got severe debilitating headaches because of this and for a number of years took narcotic pain relievers for these headaches. I also contracted Ehrlichiosis, which is a tick-borne disease (different from Lyme Disease) for which I took heavy-duty antibiotics. I then developed 2 cracked vertebrae and ended up taking some steroids. I had x-rays taken, and an MRI. One year later I developed what I thought was a bad case of sinus infection and took three different doses of antibiotics. The sinus infection got worse. I turned to homeopathic treatments for the sinus infection. I tried many different herbs and treatments to no avail. After 6 months I was so foggy I had extreme difficulty processing thoughts, no attention span at all, and was so tired I was taking at least 2 naps a day. I grew desperate and went online and realized I had Candida. One of the sympathetic ladies at my local health food store gave me Wil's phone number and in a moment of desperation I called him. At this point depression was so severe that thoughts of suicide had escalated to making plans.

My initial impression of Wil was that he was radical and maybe too radical for me. After giving it some thought and remembering that it takes radical people and ideas to facilitate change, I called him back.

I started getting treatment and the symptoms seemed to worsen at first. Then I started to notice an increase in my energy level.

I remember one time I literally dragged myself to see him barely holding tears at bay as I was feeling so fatigued and miserable. After our session my eyes were sparkling and I virtually bounced out of his office, as I felt so much better.

I feel as if Wil's information had a lot to do with turning my electrical system "back on" as I no longer experience the numbness and I feel more vibrantly alive than I have ever felt in my life.

This illness gave me an opportunity to look within and gain a level of spirituality I have never had before. I now am amazed at how much better I am feeling. I am extremely grateful for Wil, as I feel his input empowered me to facilitate my body's ability to heal itself. Wil has always been available for support dialogue, and has been a huge support for me.

> "Risk more than others think is safe. Care more than others think is wise. Dream more than others think is practical. Expect more than others think is possible."
> Cadet Maxim

Lynn D, MN

Before I met Wil I was going to the doctor at least a couple of times a month for ailments such as colds, flu, allergies, and general fatigue. I was put on a lot of antibiotics and oftentimes they did not help at all. I also had noticed a lump in my breast. I have had Wil work on me, and with the help of his diet and circuitry re-set I have not had to go to the doctor or use antibiotics, at all. Also, I now have no lumps in my breast, I am not tired all the time, and my quality of life has improved 100% all due to Wil's knowledge of the human body. His products have made me feel better too, giving me more energy and also a sense of wellbeing. I think also that my immune system has been built up, specifically because of what Wil has done for me.

Sue M, MN

I have to rave about what's happening to me now! I imagine other people feel this energy every day and take it for granted. Maybe I have in the past. But to feel this right now, it's so wonderful! I don't know if I've ever before felt this good.

I thank you and send good vibes to you and your family. That may include me now as I feel we are family.

Dixie B, Dexter, OR

Having been to the Gulf Coast of Florida to see family and friends from June until Oct. 2010, I became very ill within months.

In August I awoke to go for my usual daily walk and when I stood up my feet had lost feeling in them and I hit the floor. November my doctor informed me he couldn't do anything further for me and sent me to see an Internist.

By Dec. 7th, I was curled up in a ball making my funeral arrangements when I saw the information about the Body Electrician. When I placed the call to Wil I wasn't eating at all. I went days without food and if I tried to eat I would vomit.

When I called Wil he was much attuned to what I was dealing with. We talked at length even with his phone beeping non-stop from other callers. Wil explained the program and sent a package right out for me. I called Wil after reading the booklet and the other information he provided. We talked in length again to be sure I followed along.

The following week I was out of my wheelchair and off all

my meds. I am eating again. I have been out of the house and enjoying my life again. My family is very happy with my outcome and I am thrilled that I made the call to Wil.

RLR, FL

I recently received a Vibrational Re-Set from Wil Spencer and am delighted with his work. For years I have had pain in both hips. I attributed it to childhood trauma and later to a degenerative disc in my lower spine. Being in the healing field myself I have approached this problem holistically and always enter into a dialogue with pain or illness, and then follow up with the emotional release work that my body requires. For this chronic problem I have had acupuncture, energy work, chiropractic, and allopathic medical processes. Sometimes the pain would ease up, but it wouldn't hold. Wil gave me a Vibrational Re-Set session and I have had no hip pain since. I found Wil comfortable to be with, and his honesty impressed me. His abilities are multi-level, and I recommend a Vibrational Re-Set session with Wil to everyone.

Mir'iam C, Mountain Home, AR

After 30 years of following the alternative path of health food and wholistic health care, I always knew there was still something missing to the overall puzzle. Listening to Wil, I was struck with 'this is the missing piece!' or at least a significant piece of the puzzle. I made the dietary and supplemental changes he suggested and had a Re-Set session. After the Re-Set I had a sense of being at ease like never before. My inner tension was relieved and I have not been nearly as emotionally over sensitive, as I was in my whole adult life, since having that session. Within two months I lost 18 pounds, had increased energy, and much better sleep. My cravings for sweets are gone! I am changed.

My father also changed his eating patterns and began a more intensive supplement program than I was on. He was 82 years old, an insulin-dependent diabetic with neuropathy and was steroid-dependent after years of battling gout and rheumatoid arthritis. Dad lived with constant pain from four disintegrated discs in his lower back which prevented him from walking and required his use of a wheelchair. In six weeks Dad was completely off of insulin as well as his other diabetes medication. In three months he lost 40 pounds, his doctor took him off all other medication and his back pain was reduced to mild discomfort. His gout attacks virtually stopped and he was not suffering daily as he had been. Any improvement would have been a welcome blessing but Dad's results from the simple changes Wil suggested have been life altering.

DW, Bethlehem, PA

I have just had my third child in five years. I am 23 years old and had my first child at 18, becoming pregnant shortly after a stay in rehab for cocaine addiction. My kidneys were severely stressed and that first baby was born very early and underweight. I seizured after delivery and baby spent her first month of life in the neonatal unit. Baby 2 came three years later and in the last trimester my blood pressure began to rise steadily. Worried about my kidneys, the doctors scheduled an early induced birth, but the labor began before their intervention. Kidney problems after the birth took a few months to normalize. When I became pregnant the third time I was immediately identified as high risk and monitored constantly. Baby 3 was born three weeks ago at this writing, full term, perfectly healthy and there was absolutely no problem. In fact I was full of energy to the day of delivery; my weight gain was perfect and blood pressure stable all the way through. When I found out I was expecting this third time we immediately spoke with Wil. I got a Vibrational Re-Set to release trauma stored in cellular memory, which noticeably calmed me right away. I began adjusting my diet and adding the supplements. There was a slight kidney infection in the beginning which the doctor never treated because the supplements allowed my own body to take care of it. I believe this pregnancy would have followed suit with the other two if not for Wil and his knowledge of what it takes to be a well being. I am still following the program and breastfeeding my little girl. She did have a blistery diaper rash the other day, which Wil felt was from my Candida overload still processing out. I gave her Latero-Flora® on the tip of my finger several times and sprinkled some of the powder right on the rash. Happily, the next day the rash was gone.

KA, Bangor, PA

Wil is a great guy and very knowledgeable. He has given me a much clearer understanding than I once had of the total picture of personal health and well-being. He has helped me to become much healthier, both physically and mentally, by explaining in detail how our bodies function and what makes them work well, and providing lots of information about what foods our bodies need and what foods cause problems. He is also the first person to explain to me why emotional health is so very important.

We met through a mutual friend. When I first heard Wil's concepts regarding health, I was immediately aware that everything he was saying related directly to me and my circumstances. I was emotionally distraught, caring for a handicapped daughter, and on the verge of divorce. I seemed to hurt all over and had headaches almost daily. I just wanted to sleep all the time, and no matter how much sleep I got, it wasn't ever enough. When I found out I had cancer, I became Wil's client.

Wil worked with me quite regularly at first, energetically and physically, using his Vibrational Re-Set process. He gave me recommendations for dietary changes, and supplements to take. He also talked to me about my emotional state, and instructed me in meditation. Eventually we became good friends.

Since I began working with Wil, my emotional strength and health have greatly improved. The cancer is gone, with no medical treatment, and physically I feel much healthier and more energetic. I don't have headaches any more. All of my chronic joint pain is gone. I can get by easily on 6 hours of sleep and not feel fatigued. My skin is healthier and I have lost some weight. The ridges in my fingernails have disappeared and my Candida infection is better. I don't feel depressed, and negative thoughts are few and far between. I guess I am just much happier with me! I wake up each day thankful for the blessings I have and friends like Wil. I want to thank him for all of the help, knowledge, support, and information he has shared with me.

Lorna, MN

Notes